D0983983

STUDIES IN PARTIAL
DIFFERENTIAL EQUATIONS

MAA STUDIES IN MATHEMATICS

Published by
THE MATHEMATICAL ASSOCIATION OF AMERICA

———

Committee on Publications
E. F. BECKENBACH, Chairman

Subcommittee on MAA Studies in Mathematics
G. L. WEISS, Chairman
D. W. ANDERSON
A. C. TUCKER

Studies in Mathematics

The Mathematical Association of America

Luis A. Caffarelli
New York University, Courant Institute

David S. Jerison
Massachusetts Institute of Technology

Carlos E. Kenig
University of Minnesota, Minneapolis

Walter Littman
University of Minnesota, Minneapolis

Johannes C. C. Nitsche
University of Minnesota, Minneapolis

Steven Orey
University of Minnesota, Minneapolis

James Ralston
University of California, Los Angeles

Studies in Mathematics

Volume 23

STUDIES IN PARTIAL DIFFERENTIAL EQUATIONS

Walter Littman, editor
University of Minnesota, Minneapolis

Published and distributed by
The Mathematical Association of America

INTRODUCTION

This is a collection of articles about partial differential equations (p.d.e.) intended to be accessible to nonexperts. The articles vary somewhat in difficulty. A good background in analysis is recommended, and an elementary partial differential equations course would be helpful. The presence of a "local p.d.e. expert" to help one over an occasional hurdle would make for smoother sailing, of course.

The subject began more than two hundred years ago as a mathematical study of certain physical and geometric problems. *Laplace's equation*, for example, arose in fluid mechanics as well as in electrostatics (in the determination of the potential due to an electric charge distribution on a conducting surface). As their names indicate, the *wave equation* dealt with wave propagation problems arising in acoustics and optics, and the *heat equation* arose in the study of heat conduction problems. (For a short history of Laplace's equation and its role in the development of p.d.e., and analysis in general, see [5].)

For many years the study of these three equations constituted the bulk of research effort in p.d.e. As the more obvious questions began to be answered for these equations, mathematicians turned to generalization. For example, the natural question arose: "What classes of equations are the natural generalizations of the three basic ones, both with respect to the properties of their solutions, as

ix

well as with respect to the physical problems giving rise to them?"
This was the origin of the three "types" of equations, *elliptic*
(describing "steady state" phenomena), *hyperbolic* (describing
wave-like phenomena), and *parabolic* (describing diffusion phenomena). Until the late fifties the study of p.d.e. consisted, for all practical
purposes, of the study of these three "types" of equations. At that
time a new point of view emerged, the study of p.d.e. "independent
of type," or, p.d.e. "of general type." The philosophy here is to ask,
what is the relationship between the properties of the solutions of a
(linear) equation and the nature of the coefficients? Very often
there is an important intermediate step: "a priori inequalities." To
be more precise, certain properties of the solutions are linked to
certain inequalities that solutions of the equation or of a related
equation must satisfy. These inequalities are in turn linked to
formal relationships between the coefficients. Examples of such
properties are smoothness, local existence, and the unique continuation property. Although this theory is now one of the mainstreams
of p.d.e., perhaps the best single source is still Hörmander's book
[6]. Of course we eagerly await that author's new book on the
subject, which we understand is to appear in the near future.

Despite the emergence of the point of view just described,
research in the area of equations of the traditional types has
continued at an accelerated pace. In elliptic equations, many stub-
born problems have been solved. For accounts of this success story
see [15], [2]. For a neat exposition of higher order boundary
problems see for example [1].

During all this activity, Laplace's equation has to a large extent
been left out of the limelight. However, very recently there has been
a renewed interest in this "grandfather equation" for problems in
domains with a "rather rough" boundary. Our first article, by
Carlos Kenig and David Jerison, gives an account of some of this
recent research.

However, the minimal surface equation, the *nonlinear* elliptic
equation *par excellence*, has been very much in the limelight during
the last several decades, perhaps because of its unique position at
the crossroads between geometry and p.d.e. Our second article, by

Johannes Nitsche, makes much of the story of this equation accessible to the nonspecialist.

One of the most dramatic mathematical developments has been the relationship discovered between parabolic equations and probability theory. This connection has become a two-way bridge enriching both disciplines. Our third article by Steven Orey gives an account of this relationship.

The study of propagation of singularities of solutions of equations (especially hyperbolic) has been assuming an ever-increasing role within the subject of p.d.e. Not only is it of physical interest (for example, it tells us how light rays travel) but it has many theoretical applications. The article by J. Ralston gives us a new, essentially self-contained, treatment of this difficult subject. For a more usual approach using heavier machinery see [13]. See also [9] for additional physical motivation and background material.

The article by Caffarelli and Littman gives an "elementary" derivation for a representation formula for solutions to $\Delta u - u = 0$ in R^n. It illustrates the use of Fourier series as well as generalized functions (more general than distributions) in p.d.e.

Let us point to a few other areas not covered so far before closing this introduction. First, there is the general area of singular integrals, pseudodifferential and Fourier integral operators, which have done so much to change the map of the subject. In addition to the well-known recent books by Treves and Taylor, we refer to [3], [4], [13] for a glimpse of the power that these tools have given to the subject.

For other problems in hyperbolic equations see [11], [12]; for scattering theory [8] and [16]; for variational inequalities [10] (first two chapters); for other articles of interest, see, for example, [7] and [14].

We hope that the articles in this collection will, at least to some extent, transfer the excitement of the research process from the researcher to the reader.

Finally, it is a pleasure to thank the individual contributors for the great effort they have put into this collection.

WALTER LITTMAN

REFERENCES

1. S. Agmon, "Lectures on boundary value problems," *Van Nostrand Mathematical Studies No. 2*, 1965.
2. E. Bombieri, "Variational problems and elliptic equations," *Proc. Internat. Congress Math.*, Vancouver 1974, vol. 1, pp. 53–63.
3. A. P. Calderón, "Singular integrals," *Bull. Amer. Math. Soc.*, **72** (1966), 427–465.
4. C. Fefferman, "Recent progress in classical Fourier analysis," *Proc. Internat. Congress Math.*, Vancouver 1974, pp. 95–118 (especially 97–104).
5. L. Gårding, "The Dirichlet problem," *Math. Intelligencer*, **2** (1979/80), No. 1, 43–53.
6. L. Hörmander, *Linear Partial Differential Operators*, Springer-Verlag, 1963.
7. M. Kac, "Can one hear the shape of a drum?" *Amer. Math. Monthly*, **73**, No. 4, Part II, 1–23.
8. T. Kato, "Scattering theory," *MAA Studies in Math.*, No. 7, 1971, pp. 90–115.
9. J. B. Keller, "Rays, waves and asymptotics," *Bull. Amer. Math. Soc.*, **84** (1978), 727–750.
10. D. Kinderlehrer and G. Stampacchia, *An Introduction to Variational Inequalities and Their Applications*, Academic Press, New York, 1980.
11. H. O. Kreiss, "Initial value problems for hyperbolic partial differential equations," *Proc. Internat. Congress Math.*, Vancouver 1974, pp. 127–137.
12. C. S. Morawetz, "Nonlinear conservation equations," *Amer. Math Monthly*, **86** (1979), 284–287.
13. L. Nirenberg, "Lectures on linear partial differential equations," *Regional Conf. Series in Math.*, No. **17**.
14. ———, "Variational and topological methods in nonlinear problems," *Bull. Amer. Math. Soc.*, (N. S.) **4** (1981), 267–302.
15. J. Serrin, "The solvability of boundary value problems," *Proc. Symp. Pure Math.*, **28**, Part 2 (1976), 597–624.
16. C. H. Wilcox, "Scattering theory for the wave equation in exterior domains," *Lecture Notes in Math.*, No. **442**, Springer-Verlag, 1975.

CONTENTS

BOUNDARY VALUE PROBLEMS ON LIPSCHITZ DOMAINS

David S. Jerison* and Carlos E. Kenig*

A harmonic function u is a twice continuously differentiable function on an open subset of $\mathbb{R}^n$, $n \geqslant 2$, satisfying the Laplace equation

$$\Delta u = \left(\sum_{j=1}^{n} \frac{\partial^2}{\partial x_j^2} \right) u = 0.$$

Harmonic functions arise in many problems in mathematical physics. For example, the function measuring gravitational or electrical potential in free space is harmonic. A steady state temper-

---
*The first author was supported by an NSF post-doctoral fellowship. The second author was supported in part by the NSF. Part of this article was written while the first author was visiting the University of Minnesota, whose hospitality he acknowledges.

ature distribution in a homogeneous medium also satisfies the Laplace equation.

We will be concerned with the two basic boundary value problems for the Laplace equation, the Dirichlet and Neumann problems. Let D be a bounded, smooth domain in $\mathbb{R}^n$ and let f be a smooth (i.e., C^∞) function on ∂D, the boundary of D. The *Dirichlet problem* is to find (and describe) a function u that is harmonic in D, continuous in $\overline{D}$, and equals f on ∂D. This corresponds to the problem of finding the temperature inside a body D when one knows the temperature f on ∂D. The *Neumann problem* is to find a function u that is harmonic in D, belongs to $C^1(\overline{D})$, and satisfies $\partial u/\partial N = f$ on ∂D, where $\partial u/\partial N$ represents the normal derivative of u on ∂D. This corresponds to the problem of finding the temperature inside D when one knows the heat flow f through the boundary surface ∂D.

Our main purpose here is to describe results on the boundary behavior of u in the case of smooth domains and the extension of these results to the case of domains with corners (Lipschitz domains, Definition (1.27)). The boundaries of these domains have the borderline amount of smoothness necessary for the validity of theorems like the one stated below.

In a smooth domain, the method of layer potentials yields the existence of a solution u to the Dirichlet problem with boundary data $f \in C^{k,\alpha}(\partial D)$ and the bound

$$\|u\|_{C^{k,\alpha}(\overline{D})} \leqslant A_{k,\alpha}\|f\|_{C^{k,\alpha}(\partial D)}{}^\dagger, \qquad k = 0,1,2,\dots.$$

$$0 < \alpha < 1$$

(Uniqueness of u follows immediately from the maximum principle.) In certain bad (nonsmooth) domains a solution for continuous boundary data f need not exist. The problem of describing these domains was settled completely by N. Wiener. (See Section 1.)

---

† The $C^{k,\alpha}$ norm is the supremum of all derivatives up to order k plus

$$\sup_{\substack{x,y \in \partial D \\ |\beta| = k}} \frac{|D^\beta f(x) - D^\beta f(y)|}{|x-y|^\alpha}.$$

What happens if the size of f is measured in the L^2 norm? This is of interest as a measure of the variation in data even if we are only concerned with continuous functions: If $f_1 - f_2$ has small L^2 norm, we want to know that the corresponding solutions u_1 and u_2 are near each other. The wisdom of hindsight tells us that as long as we are going to examine all continuous functions in L^2 norm, it is no harder to consider arbitrary functions in L^2. Another reason to consider the L^2 norm is that it is better suited to the Neumann problem than C^k norms, even on smooth domains. Our first task is to formulate an appropriate theorem by examining a model case, namely the unit ball B. The first person to consider this sort of question was Fatou [19], who examined the case of the unit disc in $\mathbb{R}^2$ with $f \in L^\infty$ rather than L^2. In the first section we will prove the following theorem of Fatou type. Let $d\sigma$ denote surface measure of ∂B.

THEOREM. *Suppose that* $1 < p \leqslant \infty$ *and* $f \in L^p(\partial B, d\sigma)$. *Then there exists a unique harmonic function u in B such that* $\lim_{r \to 1} u(rQ) = f(Q)$ *for almost every* $Q \in \partial B$, *and*

$$\int_{\partial B} u^*(Q)^p \, d\sigma(Q) \leqslant C_p \int_{\partial B} |f(Q)|^p \, d\sigma(Q), \qquad (*)$$

where $u^*(Q) = \sup_{0 \leqslant r < 1} |u(rQ)|$.

The theorem asserts that $f_r(Q) = u(rQ)$ converges to $f(Q)$ not only in L^p norm, but also in the sense of Lebesgue's dominated convergence. (In the analogous estimates to $(*)$ in the Neumann problem, u is replaced by the gradient of u. In that case the estimate fails for $p = \infty$, even if $\partial u / \partial N$ is continuous.)

There is an appropriate endpoint result at $p = 1$ with $L^1(\partial B, d\sigma)$ replaced by finite measures on ∂B. The result puts positive harmonic functions in B in one-to-one correspondence with positive measures on ∂B. In particular,

COROLLARY. *Every positive harmonic function u in B has a finite radial limit* $\lim_{r \to 1} u(rQ)$ *for almost every* $Q \in \partial B$.

In the theorem and its corollary, the radial limit can be replaced by a *nontangential limit*: if X tends to Q with $|X - Q| < (1 + \alpha)\text{dist}(X, \partial B)$ for some fixed $\alpha > 0$, then $u(X)$ has a limit for almost every Q.

The nature of the solution u to the Dirichlet problem changes as the domain becomes less smooth. This change is reflected in the need for alternative techniques to solve for u, but is best described in terms of a notion called harmonic measure. Let D be a bounded Lipschitz domain. As we shall see in the first section, Lipschitz domains are among the domains for which the solution to the Dirichlet problem exists for any $f \in C^0(\partial D)$. Given $X \in D$, the mapping $f \mapsto u(X)$ is a continuous linear functional on $C^0(\partial D)$. Therefore, by the Riesz representation theorem, there is a unique Borel measure ω^X on ∂D such that

$$u(X) = \int_{\partial D} f(Q) \, d\omega^X(Q).$$

ω^X is called *harmonic measure* for D evaluated at X. For example, harmonic measure for B evaluated at the origin is a constant multiple of surface measure: $\omega^0 = \sigma/\sigma(\partial B)$ (the mean value Theorem (1.6)).

Fix $X_0 \in D$, and denote $\omega = \omega^{X_0}$. The importance of harmonic measure to the boundary behavior of harmonic functions can be illustrated by the following theorem. If u is a positive harmonic function, then u has finite nontangential limits almost everywhere with respect to ω (see the corollary above, L. Carleson [9], and R. Hunt and R. Wheeden [24]). Conversely, given any set $E \subset \partial D$ with $\omega(E) = 0$, there is a positive harmonic function u in D with $\lim u(X) = \infty$ as $X \to Q$ for every $Q \in E$.

The difficulty with harmonic measure is that it is hard to calculate explicitly. In general, harmonic measure may be very different from surface measure. If D is a $C^{1,\alpha}$ domain (see (1.27)), then harmonic measure and surface measure are essentially identical in that each is a bounded multiple of the other. This can be proved by the classical method of layer potentials. Along the same lines, one can use layer potentials to solve the Dirichlet and Neumann problems with boundary data in L^p. On C^1 domains (see

(1.27)), it is no longer true that harmonic measure is a bounded multiple of surface measure or vice versa. Moreover, the tame singularity of the layer potential is transformed into one that is unmanageable by classical means. However, recently B. E. J. Dahlberg [12] proved that on a C^1 or even a Lipschitz domain harmonic measure and surface measure are mutually absolutely continuous. Using a quantitative version of mutual absolute continuity, he proved that in a Lipschitz domain D one can solve the Dirichlet problem as in the theorem above with $f \in L^2(\partial D, d\sigma)$. (For $f \in L^p(\partial D, d\sigma)$ with $p < 2$ this can fail!) In Section 3 we present a proof of Dahlberg's theorem given in [26] based on an elementary integral formula.

A shortcoming of Dahlberg's method was that it was not applicable to the Neumann problem. In an independent approach, A. P. Calderón, et al., [8] showed, using a deep result of A. P. Calderón [7], that the method of layer potentials was applicable to C^1 domains after all. They deduced sharp estimates as in (*) in the theorem above in the Dirichlet and Neumann problems for boundary data in L^p. (On C^1 domains, the theorems are valid in the full range of values of p.) Finally, Jerison and Kenig [29] proved sharp estimates in the Neumann problem on Lipschitz domains for boundary data in L^2. They used another elementary integral formula and Dahlberg's theorem (Section 4). The integral formula is related to Green's formulas and has been used by Payne and Weinberger [40] and F. Rellich [41]. Further bibliographical comments can be found at the end of the article in a section called "Notes."

While estimates like (*) of the theorem above are the right ones from the point of view of the harmonic analyst, they may seem foreign to the reader who is familiar with other approaches to partial differential equations. For that reason we would like to point out that one can use them in a straightforward manner to deduce sharp estimates in the Dirichlet and Neumann problems in terms of Sobolev spaces for which no other proofs are known (see 6.3). Moreover, all the estimates above are valid when the Laplace operator is replaced by a second order uniformly elliptic divergence class operator with smooth coefficients.

1. BACKGROUND: THE CLASSICAL DIRICHLET PROBLEM

Throughout this article D will denote a bounded, connected, open subset of $\mathbb{R}^n$, $n \geq 2$. This section is an outline of some of the basic facts and methods of potential theory. Only a few detailed proofs are given. For the rest, the reader is referred to the works mentioned in the notes.

By the classical Dirichlet problem we mean

(1.1) *Given $f \in C^0(\partial D)$, find u harmonic in D, continuous in $\overline{D}$ such that $u|_{\partial D} = f$.*

As a consequence of the (weak) maximum principle (1.11), if such a u exists, it is necessarily unique. However, u need not exist. Domains with the property that u does exist for every $f \in C^0(\partial D)$ are called *regular* for the Dirichlet problem (see (1.22)).

The classical potential theory that we will describe is based on three explicit calculations. First and foremost is the *fundamental solution F* for Laplace's equation. $\Delta F = \delta$, where δ is the unit point mass at the origin in $\mathbb{R}^n$. There are many choices for F differing from each other by global harmonic functions on $\mathbb{R}^n$. Since the Laplace operator is rotation invariant, it is reasonable to expect that one solution depends only on $|X|$. Indeed, ([21])

$$F(X) = \begin{cases} \dfrac{-1}{(n-2)\omega_n |X|^{n-2}} & n > 2 \\[2mm] \dfrac{1}{2\pi} \log|X| & n = 2 \end{cases} \tag{1.2}$$

where ω_n is the surface measure of the unit sphere in $\mathbb{R}^n$. $F(x)$ is the electrical potential in free space induced by a unit charge at the origin. It provides a formula for a solution w to the equation $\Delta w = \psi$, where $\psi \in C_0^\infty(\mathbb{R}^n)$:

$$w(X) = F * \psi(X) = \int_{\mathbb{R}^n} F(X-Y)\psi(Y)\,dY.$$

The formal justification is that

$$\Delta(F * \psi) = (\Delta F) * \psi = \delta * \psi = \psi.$$

The second calculation is one that describes completely the solution to the Dirichlet problem in the unit ball B in $\mathbb{R}^n$:

$$u(X) = \int_{\partial B} P(X, Q) f(Q) \, d\sigma(Q),$$

where

$$P(X, Q) = \frac{1 - |X|^2}{\omega_n |X - Q|^n}. \tag{1.3}$$

P is known as the *Poisson kernel*. u is the *Poisson integral* of f. We will derive (1.3) using (1.2) shortly. We will then use it to examine the Dirichlet problem in the ball in detail.

The third calculation is that of a particular harmonic function in a right circular cone. It will enable us to give a simple sufficient condition for D to be regular, called the exterior cone condition (1.24).

Green's formulas. Let us recall some results from advanced calculus.

(1.4) THE DIVERGENCE THEOREM. *Let D be a smooth domain and $H = (h_1, \ldots, h_n)$ a vector-valued function in $C^1(\overline{D})$. Then*

$$\int_D \operatorname{div} H = \int_{\partial D} \langle N_Q, H(Q) \rangle \, d\sigma(Q)$$

where

$$\operatorname{div} H(X) = \sum_{j=1}^{n} \frac{\partial h_j}{\partial x_j}(X),$$

N_Q is the outer unit normal to ∂D at Q, $d\sigma$ is surface measure on ∂D, and $\langle , \rangle$ is the inner product. (In integrals of volume such as $\int_D \operatorname{div} H$, the volume element dX will be omitted unless the variable of integration is ambiguous.)

A corollary of the divergence theorem is

(1.5) GREEN'S FORMULAS. *If D is a smooth domain and u and v
belong to $C^2(\overline{D})$, then*

$$\int_D u \Delta v = - \int_D \langle \nabla u, \nabla v \rangle + \int_{\partial D} u \frac{\partial v}{\partial N} \, d\sigma,$$

$$\int_D (u \Delta v - v \Delta u) = \int_{\partial D} \left(u \frac{\partial v}{\partial N} - v \frac{\partial u}{\partial N} \right) d\sigma.$$

$\partial/\partial N = \langle N_Q, \nabla \rangle$ *denotes differentiation in the unit outer normal
direction.*

The second formula with $v \equiv 1$ implies

(1.6) THE MEAN VALUE THEOREM. *Let u be harmonic in D,
$X \in D$; $B(X, r) = \{P \in \mathbb{R}^n : |P - X| < r\}$ and $\overline{B(X, r)} \subset D$. Then*

$$u(X) = \frac{1}{\sigma(\partial B(X, r))} \int_{\partial B(X, r)} u(Q) \, d\sigma(Q)$$

$$= \frac{1}{\omega_n} \int_{\partial B} u(X + rQ) \, d\sigma(Q),$$

where B is the unit ball $B = \{P \in \mathbb{R}^n : |P| < 1\}$; $\omega_n = \sigma(\partial B)$.

(1.7) CONVERSE OF THE MEAN VALUE THEOREM: *Suppose that u
is continuous in D and whenever $\overline{B(X, r)} \subset D$ we have $u(X) = \frac{1}{\omega_n} \int_{\partial B} u(X + rQ) \, d\sigma(Q)$. Then $u \in C^\infty(D)$ and u is harmonic in D.*

(1.8) COROLLARY. *If u is harmonic in D, then $u \in C^\infty(D)$.*

(1.9) COROLLARY. *If $\{u_k\}$ is a sequence of harmonic functions in
D that converges uniformly on compact subsets of D to a limit u, then
u is harmonic in D. Also any derivative of $\{u_k\}$ converges uniformly
on compact subsets to the corresponding derivative of u.*

(1.10) THE MAXIMUM PRINCIPLE. *If u is harmonic and real-valued in an open, connected set D and* $\sup\{u(X): X \in D\} = A < \infty$, *then either* $u(X) < A$ *for all* $X \in D$ *or* $u(X) \equiv A$ *for all* $X \in D$.

(1.11) COROLLARY (WEAK MAXIMUM PRINCIPLE). *If u is continuous on the compact set* $\overline{D}$ *and harmonic in D, then the maximum value of u on* $\overline{D}$ *is achieved on* ∂D.

It will be convenient to put $F(X, Y) = F(X - Y)$. Notice that $\Delta_Y F(X, Y) = \delta(X - Y)$. The fundamental solution in a bounded domain is known as the *Green function* $G(X, Y)$. It is the function on $\overline{D} \times \overline{D}$, continuous for $X \neq Y$ and satisfying

$$\Delta_Y G(X, Y) = \delta(X - Y), \qquad X \in D;$$

$$G(X, Y) = 0, \qquad X \in D, \qquad Y \in \partial D. \qquad (1.12)$$

$G(X, Y)$ as a function of Y is the potential induced by a unit charge at X that is grounded to zero potential on ∂D. The Green function has a formula in terms of the solution to the Dirichlet problem (1.1). Let $u_X(Y)$ be the harmonic function with boundary values $u_X(Y)|_{\partial D} = F(X, Y)|_{\partial D}$. Then from (1.12),

$$G(X, Y) = F(X, Y) - u_X(Y). \qquad (1.13)$$

Thus the Green function exists if (1.1) can be solved.

Green's approach was to reverse this point of view and write the solution of (1.1) in terms of $G(X, Y)$. The formal reasoning is that if $\Delta u = 0$, then using (1.5) and (1.12),

$$u(X) = \int_D u(Y)\delta(X - Y)\, dY = \int_D u(Y)\Delta_Y G(X, Y)\, dY$$

$$= \int_D \left[u(Y)\Delta_Y G(X, Y) - (\Delta u(Y))G(X, Y) \right] dY$$

$$= \int_{\partial D} \left[u(Q)\frac{\partial}{\partial N_Q} G(X, Q) - \frac{\partial u}{\partial N_Q}(Q)G(X, Q) \right] d\sigma(Q)$$

$$= \int_{\partial D} u(Q)\frac{\partial}{\partial N_Q} G(X, Q)\, d\sigma(Q).$$

(Subscripts Y and Q indicate the variable in which differentiation occurs.)

Thus we have derived the formula

$$u(X) = \int_{\partial D} f(Q) \frac{\partial}{\partial N_Q} G(X, Q) \, d\sigma(Q) \qquad (1.14)$$

for the harmonic function u with boundary values f. (One can show that $G(X, Y) = G(Y, X)$, so that the function $\frac{\partial}{\partial N_Q} G(X, Q)$ is harmonic as a function of X.)

Formula (1.14) is valid for a large class of domains—even Lipschitz domains. We will use it for the far simpler task of calculating the Poisson kernel for the ball, (1.3). Recognizing the symmetry of the fundamental solution $F(X)$ under the inversion $X \to X/|X|^2$, we obtain Green's function for the unit ball, when $n > 2$:

$$G(X, Y) = \begin{cases} F(X, Y) - |X|^{2-n} F(X/|X|^2, Y), & X \neq 0 \\ F(0, Y) - (2-n)^{-1} \omega_n^{-1}, & X = 0. \end{cases}$$

Hence the Poisson kernel must be

$$P(X, Q) = \frac{\partial}{\partial N_Q} G(X, Q) = \frac{1 - |X|^2}{\omega_n |X - Q|^n}.$$

The Dirichlet Problem in the Ball.

(1.15) THEOREM. *If u is the Poisson integral of $f \in C^0(\partial B)$ as in (1.3), then u is harmonic in B, continuous in $\overline{B}$, and $u|_{\partial B} = f$.*

Proof. u is harmonic in B because, by direct calculation, $P(X, Q)$ is harmonic as a function of X. We claim that

(a) $\int_{\partial B} P(X, Q) \, d\sigma(Q) = 1$ for $X \in B$.
(b) For $Q \in \partial B$ and any neighborhood V of Q,

$$\lim_{r \to 1} \int_{\partial B \setminus V} P(rQ, Q') \, d\sigma(Q') = 0.$$

(b) follows from inspection of (1.3). To prove (a), notice that the mean value theorem (1.6) implies that for $0 \leqslant r < 1$, $Q' \in \partial B$,

$$1 = \omega_n P(0, Q') = \int_{\partial B} P(rQ, Q') \, d\sigma(Q).$$

But $P(rQ, Q') = P(rQ', Q)$, so setting $X = rQ'$, we obtain (a).

Given $\varepsilon > 0$, choose $\delta > 0$ such that $|Q - Q'| < \delta$ implies $|f(Q) - f(Q')| < \varepsilon$. (This is possible because ∂B is compact.) Denote $V(Q) = \{Q' \in \partial B : |Q - Q'| < \delta\}$. Let Q_0, Q_1 belong to ∂B, $X = rQ_1$, and $|X - Q_0| < \delta/M$, $M \geqslant 1$.

$$|u(X) - f(Q_0)| \leqslant |u(X) - f(Q_1)| + |f(Q_1) - f(Q_0)|$$

$$\leqslant \left| \int P(X, Q)(f(Q) - f(Q_1)) \, d\sigma(Q) \right| + \varepsilon$$

$$\leqslant \left| \int_{V(Q_1)} P(X, Q) \varepsilon \, d\sigma(Q) \right|$$

$$+ 2 \sup |f| \int_{\partial B \setminus V(Q_1)} P(rQ_1, Q) \, d\sigma(Q) + \varepsilon$$

$$\leqslant 2\varepsilon + 2 \sup |f| \int_{\partial B \setminus V(Q_1)} P(rQ_1, Q) \, d\sigma(Q)$$

by (a). For M sufficiently large, r is very close to 1, and hence by (b), the final term is dominated by ε. Thus we have shown simultaneously that u is continuous at each point of ∂B and that its limit is f.

(1.16) THEOREM. *Suppose that $1 \leqslant p \leqslant \infty$ and $f \in L^p(\partial B, d\sigma)$. Let u be the Poisson integral of f. Then $\lim_{r \to 1} u(rQ) = f(Q)$ for almost every $Q \in \partial B, d\sigma$. Moreover, if $u^*(Q) = \sup_{0 \leqslant r < 1} |u(rQ)|$, then*

(a) $\|u^*\|_{L^p(\partial B, d\sigma)} \leqslant C_p \|f\|_{L^p(\partial B, d\sigma)}, \ 1 < p \leqslant \infty$;

(b) $\sigma\{Q \in \partial B : u^*(Q) > \lambda\} \leqslant \dfrac{C}{\lambda} \|f\|_{L^1(\partial B, d\sigma)}.$

Proof. The heart of the matter is (b). There is a well-known argument using (b) and (1.15) that shows $u(rQ)$ tends to $f(Q)$ when $f \in L^1(\partial B, d\sigma)$. This argument will be presented later (2.3′). Denote

$$\Delta = \Delta(Q_0, s) = \{Q \in \partial B: |Q - Q_0| < s\};$$

$$\mathfrak{M}_\sigma f(Q) = \sup\left\{\frac{1}{\sigma(\Delta)} \int_\Delta |f|\, d\sigma: Q \in \Delta\right\}.$$

The estimates (a) and (b) with u^* replaced by $\mathfrak{M}_\sigma f$ are well known [44]. Therefore, it is enough to show $u^*(Q) \leqslant C\mathfrak{M}_\sigma f(Q)$. Denote

$$\Delta_j = \{Q: |Q - Q_0| < 2^j(1 - r)\}, j = 0, 1, \ldots;$$

$$A_j = \Delta_j \backslash \Delta_{j-1}, j \geqslant 1;$$

$$|u(rQ_0)| \leqslant \int_{\Delta_0} P(rQ_0, Q)|f(Q)|\, d\sigma(Q)$$

$$+ \sum_{j=1}^\infty \int_{A_j} P(rQ_0, Q)|f(Q)|\, d\sigma(Q)$$

$$\leqslant \frac{C}{(1-r)^{n-1}} \int_{\Delta_0} |f(Q)|\, d\sigma(Q)$$

$$+ \sum_{j=1}^\infty 2^{-j} \frac{C}{(2^j(1-r))^{n-1}} \int_{\Delta_j} |f(Q)|\, d\sigma(Q)$$

$$\leqslant C \sum_{j=1}^\infty 2^{-j} \mathfrak{M}_\sigma f(Q_0) \leqslant C\mathfrak{M}_\sigma f(Q_0),$$

since $P(rQ_0, Q) \leqslant C\dfrac{1-r}{(2^j(1-r))^n}$ for $Q \in A_j$.

The explicit Poisson integral formula can of course be transplanted by translation and dilation to any ball B in $\mathbb{R}^n$. As a consequence of the explicit Poisson integral formula in a ball one can prove

(1.17) *Harnack's Inequality.* Let K be a compact subset of D. There is a positive constant M depending only on K and D such that for all X, Y in K,

$$M^{-1}u(Y) \leqslant u(X) \leqslant Mu(Y)$$

for every positive harmonic function u in D.

The Perron Process. A distribution w defined in an open set D is said to be *subharmonic* if $\phi \geqslant 0$ implies $(w, \Delta\phi) \geqslant 0$ for every $\phi \in C_0^\infty(D)$. $(,)$ denotes the usual pairing. There is a representative of the distribution w that is an upper semicontinuous function. Henceforth we will consider that representative only and denote it w. Let us state without proof some further well-known properties of subharmonic functions.

(1.18) Let $w \in C^2(D)$. Then w is subharmonic if and only if $\Delta w \geqslant 0$ in D.

(1.19) *Maximum principle.* Let w be subharmonic in D and suppose that $\limsup_{X \to Q} w(X) \leqslant M$ for all $Q \in \partial D$; then either $w \equiv M$ or $w < M$ in D. (D is connected.)

(1.20) THEOREM. *Let w be subharmonic in D, then $\int_K |u(X)| \, dX < \infty$ for any compact set $K \subset D$ and $\int_{\partial B} |w(X)| \, d\sigma(X) < \infty$ for any ball B, where $\bar{B} \subset D$. Moreover, $w \leqslant P_B^w$ in B, where P_B^w is the Poisson integral in B of $w|_{\partial B}$.*

(1.21) COROLLARY. *u is harmonic in D if and only if u and $-u$ are subharmonic in D.*

$$\text{Denote } M_B^w(X) = \begin{cases} P_B^w(X) & X \in B, \\ w(X) & X \in D \setminus B. \end{cases}$$

If w is subharmonic in D, then M_B^w is subharmonic and $M_B^w \geqslant w$.

A nonempty family $\mathscr{P}$ of functions subharmonic in D is called a *Perron family* if

(a) If $w_1, w_2 \in \mathscr{P}$, then $\max\{w_1, w_2\} \in \mathscr{P}$;
(b) If $w \in \mathscr{P}$ and $\bar{B} \subset D$, then $M_B^w \in \mathscr{P}$.

(1.22) THEOREM. *Let $\mathscr{P}$ be a Perron family (D is connected), $w_0(X) = \sup\{w(X): w \in \mathscr{P}\}$; then either $w_0 \equiv +\infty$ in D or w_0 is harmonic.*

We are now ready to describe the Perron method. Let h be a continuous real-valued function on ∂D. Let $\mathscr{P}(h)$ be the family of all subharmonic functions w in D satisfying $\limsup_{x \to Q} w(X) \leqslant h(Q)$ for all $Q \in \partial D$. Denote $M = \sup\{|h(Q)|: Q \in \partial D\}$ and

$$H_f(X) = \sup\{w(X): w \in \mathscr{P}(h)\}.$$

It is clear that $\mathscr{P}(h)$ is nonempty because $-M \in \mathscr{P}(h)$. Also, $H_f(X) \leqslant M$ by the maximum principle (1.19). Thus, H_f is harmonic. If the solution to the classical Dirichlet problem for f exists, then it is H_f. Another candidate for the solution is the object obtained by the same analysis as above with all the inequalities reversed, namely $-H_{-f}$. ($-H_{-f}$ can also be written as an infimum of super-harmonic functions.)

(1.23) THEOREM. *For any bounded domain D, $H_f = -H_{-f}$ for all continuous real valued functions f.*

The domain D is called *regular* at $Q \in \partial D$ if H_f is continuous at Q for all continuous f. The domain is regular for the Dirichlet problem if it is regular at every $Q \in \partial D$. A point Q in ∂D has a *barrier function* if for some open ball $B(Q, r_0)$ there exists a subharmonic function in $B(Q, r_0) \cap D$ such that $\lim_{X \to Q} w(X) = 0$ and $\sup\{w(X): r < |X - Q| < r_0\} < 0$ for $0 < r < r_0$. An easy consequence of (1.23) and the maximum principle (1.19) is

(1.24) COROLLARY. *A bounded domain $D \subset \mathbb{R}^n$ is regular at $Q \in \partial D$ if and only if Q has a barrier function (relative to D).*

It is clear from the definition of a barrier function that if $B(Q, r) \cap D' \subset B(Q, r) \cap D$, $Q \in \partial D'$, $Q \in \partial D$, and D is regular at Q, then D' is regular at Q. N. Wiener has characterized regular points in terms of the size of $^C D$ near Q [46]. We will only need a simpler sufficient condition for the domain D to be regular.

(1.25) D satisfies the *exterior cone condition* if for every Q there exists an open cone $\mathcal{C}$ with vertex at Q such that for some $r > 0$, $\mathcal{C} \cap B(Q, r) \subset {}^C D$.

(1.26) PROPOSITION. *Domains satisfying the exterior cone condition at every boundary point are regular for the Dirichlet problem.*

Proof. We must construct a barrier function in the exterior of an arbitrarily small right circular cone. We may as well rotate the cone so that its axis is along the negative x_1 axis and its vertex is at the origin. Let $\theta = \arccos(x_1 / |X|)$. We will seek a function of the form $w(X) = |X|^2 \psi(\theta)$ defined in

$$D = \{ X = (x_1, \ldots, x_n) = x_1 > \cos(\pi - \delta)|X| \}.$$

Let

$$\phi(\theta) = \int_0^\theta \frac{1}{(\sin t)^{n-2}} \int_0^t (\sin \tau)^{n-2} \, d\tau \, dt, \, 0 \leqslant \theta < \pi.$$

Then

$$\phi''(\theta) + (n-2)(\cot \theta) \phi'(\theta) = 0.$$

Since ϕ is continuous, for any $\delta > 0$ there exist A, $\alpha > 0$ such that for $0 \leqslant \theta \leqslant \pi - \delta$,

$$\phi(\theta) - A < -1 \text{ and } 1 + \alpha(\alpha + n - 2)(\phi(\theta) - A) > 0.$$

Define $w(X) = |X|^\alpha (\phi(\theta) - A)$. Then the differential equation for ϕ implies

$$\Delta w(X) = |X|^{\alpha-2}(1 + \alpha(\alpha + n - 2)(\phi(\theta) - A)) > 0$$

in D, and w is a barrier function for D at the origin.

(1.27) *Definition.* A bounded domain D is called a *Lipschitz domain* with Lipschitz constant less than or equal to M if for any $Q \in \partial D$ there is a ball B with center at Q, a coordinate system (isometric to the usual coordinate system) $x' = (x_1, \ldots, x_{n-1})$, x_n with origin at Q and a function ϕ: $\mathbb{R}^{n-1} \to \mathbb{R}$ such that $\phi(0) = 0$, $|\phi(x') - \phi(y')| \leq M|x' - y'|$, and

$$D \cap B = \{X = (x', x_n) : x_n > \phi(x')\} \cap B.$$

If for each Q, the function ϕ can be chosen in $C^1(\mathbb{R}^{n-1})$, then D is called a C^1 *domain*. If, in addition, $\nabla\phi$ satisfies a Hölder condition of order α,

$$|\nabla\phi(x') - \nabla\phi(y')| \leq C|x' - y'|^\alpha,$$

we call D a $C^{1,\alpha}$ *domain.*

Notice that the cone

$$\mathcal{C} = \{(x', x_n) : x_n < -M|x'|\}$$

satisfies $\mathcal{C} \cap B \subset {}^C D$. Thus Lipschitz domains satisfy the exterior cone condition.

The function ϕ satisfying the Lipschitz condition $|\phi(x') - \phi(y')| \leq M|x' - y'|$ is differentiable almost everywhere and $\nabla\phi \in L^\infty(\mathbb{R}^{n-1})$. Surface measure σ is defined for each Borel subset $E \subset \partial D \cap B$ by

$$\sigma(E) = \int_{E_*} \left(1 + |\nabla\phi(x')|^2\right)^{1/2} dx',$$

where

$$E_* = \{x' : (x', \phi(x')) \in E\}.$$

The unit outer normal to ∂D given in the coordinate system by $(\nabla\phi(x'), -1)/(1 + |\nabla\phi(x')|^2)^{1/2}$ exists for almost every x'. The unit normal at Q will be denoted N_Q. It exists for almost every $Q \in \partial D$ with respect to σ.

The Method of Layer Potentials. In light of formulas (1.13) and (1.14), C. Neumann proposed the formula

$$w(X) = \int_{\partial D} f(Q) \frac{\partial F}{\partial N_Q}(X, Q) \, d\sigma(Q)$$

$$= \omega_n^{-1} \int_{\partial D} f(Q) \frac{\langle X - Q, N_Q \rangle}{|X - Q|^n} \, d\sigma(Q)$$

as a first approximation to the solution of the Dirichlet problem, $\Delta u = 0$ in D, $u|_{\partial D} = f$. $w(X)$ is known as the *double layer potential* of f. The guess is not so far off on a smooth domain D for continuous functions f. First of all, w is harmonic. Second, as $X \to Q$, $w(X) \to \frac{1}{2} f(Q) + Kf(Q)$, where K is the operator on ∂D given by

$$Kf(Q) = \omega_n^{-1} \int_{\partial D} f(P) \frac{\langle Q - P, N_P \rangle}{|Q - P|^n} \, d\sigma(P).$$

If Kf were zero, we would be done, and in some sense it is true that Kf is small compared to f. The operator $K: C^0(\partial D) \to C^0(\partial D)$ is compact. Therefore, by a general procedure due largely to Fredholm, the operator $T = \frac{1}{2}I + K$ is invertible modulo a finite dimensional subspace of $C^0(\partial D)$. If D and ^CD are connected, T is actually invertible on $C^0(\partial D)$. Therefore, the solution to the Dirichlet problem may be written

$$u(X) = \omega_n^{-1} \int_{\partial D} g(Q) \frac{\langle X - Q, N_Q \rangle}{|X - Q|^n} \, d\sigma(Q),$$

where $g = T^{-1}(f)$. The operator K is compact on $C^0(\partial D)$ even on $C^{1, \alpha}$ domains, so this procedure solves the classical Dirichlet problem in that case, too. Moreover, if D is a C^∞ domain, it is possible to show that K is compact as a mapping of $C^{k, \alpha}(\partial D)$ to $C^{k, \alpha}(\partial D)$. Hence, we can conclude that if $f \in C^\infty(\partial D)$, then $u \in C^\infty(\overline{D})$.

This approach can also be used to obtain results like Theorem (1.16) for $C^{1, \alpha}$ domains. Denote a nontangential approach region to

Q as follows:

$$\Gamma_\alpha(Q) = \{ X \in D : |X - Q| < (1 + \alpha)\text{dist}(X, \partial D) \}.$$

The *nontangential maximal function* of w is

$$N_\alpha(w)(Q) = \sup\{|w(X)| : X \in \Gamma_\alpha(Q)\}.$$

The key ingredients of the proof of the analogue of (1.16) are these. First, K is compact as a mapping from $L^p(\partial D)$ to $L^p(\partial D)$. Second,

$$N_\alpha(w)(Q) \leqslant C(\mathfrak{M}_\sigma(f)(Q) + \mathfrak{M}_\sigma(Kf)(Q)),$$

where $\mathfrak{M}_\sigma$ denotes the Hardy-Littlewood maximal operator as in the proof of (1.16) and w is the double layer potential of f. Finally, from the L^p bounds for $\mathfrak{M}_\sigma$, K, and $T^{-1} = (\frac{1}{2}I + K)^{-1}$, one obtains

(1.27) THEOREM. *Let D be a $C^{1,\alpha}$ domain, $1 < p < \infty$. If $f \in L^p(\partial D, d\sigma)$ and*

$$u(X) = \omega_n^{-1} \int_{\partial D} \frac{\langle X - Q, N_Q \rangle}{|X - Q|^n} T^{-1}f(Q)\, d\sigma(Q),$$

then $\|N_\beta(u)\|_{L^p(d\sigma)} \leqslant C_p \|f\|_{L^p(d\sigma)}$ and the harmonic function u tends to f nontangentially, i.e., $\lim_{X \to Q} u(X) = f(Q)$ for $X \in \Gamma_\beta(Q)$. (Any value of β is allowed.)

The difficulty with C^1 domains and Lipschitz domains is that even the boundedness much less the compactness of K on L^p is in doubt. However, recently A. P. Calderón [7] proved that K is bounded on C^1 domains in dimension 2. Fabes, Jodeit, and Rivière [16] extended this to n dimensions by the familiar method of rotations. In addition, they succeeded in showing that K is compact on $L^p(\partial D)$, $1 < p < \infty$, for C^1 domains. They were thus able to extend Theorem (1.27) to C^1 domains. Whether the method of layer potentials is applicable to Lipschitz domains is an open problem. A different approach will be used in Sections 3 and 4.

Let us now turn to the Neumann problem. Let D be a C^∞ domain. We seek to solve $\Delta u = 0$ in D, $\partial u / \partial N = f$, where, say, $f \in C^\infty(\partial D)$. Green's formula (1.5) with $v \equiv 1$ implies that we must have $\int_{\partial D} f \, d\sigma = 0$. When D and $^C D$ are connected, this is the only compatibility condition needed. We will consider that case only for simplicity. A good first guess at the solution u is the so-called *single layer potential of f* given by

$$v(X) = \int_{\partial D} f(Q) F(X, Q) \, d\sigma(Q).$$

Once again v is harmonic in D, and

$$\frac{\partial v}{\partial N_Q}(Q) = \frac{1}{2} f(Q) - K^* f(Q),$$

where K^* is the adjoint of K above, i.e.,

$$K^* f(P) = \omega_n^{-1} \int_{\partial D} \frac{\langle P - Q, N_P \rangle}{|P - Q|^n} f(Q) \, d\sigma(Q).$$

Thus K^* is compact from $C^{k,\alpha}(\partial D)$ to $C^{k,\alpha}(\partial D)$. Fredholm's theory says that $\tilde{T} = \frac{1}{2} I - K^*$ is invertible on the subspace of $C^{k,\alpha}(\partial D)$ of functions of mean value zero ($\int_{\partial D} f \, d\sigma = 0$). Therefore the solution to the Neumann problem can be written

$$u(X) = \int_{\partial D} (\tilde{T}^{-1} f)(Q) F(X, Q) \, d\sigma(Q)$$

and u belongs to $C^\infty(\overline{D})$. As before, $\tilde{T}$ is also invertible on $L^p(\partial D)$ with mean value zero $1 < p < \infty$. Therefore,

(1.29) THEOREM. *Let D be a $C^{1,\alpha}$ domain; (D and $^C D$ are connected). Let $1 < p < \infty$. Assume that $f \in L^p(\partial D, d\sigma)$ and $\int_{\partial D} f \, d\sigma = 0$. Then*

$$u(X) = \int_{\partial D} (\tilde{T}^{-1} f)(Q) F(X, Q) \, d\sigma(Q)$$

is harmonic in D, $\nabla u(X)$ has nontangential limits $\nabla u(Q)$ a.e. $Q \in \partial D$,
$f(Q) = \langle N_Q, \nabla u(Q) \rangle$, *and*

$$\|N_\beta(\nabla u)\|_{L^p(d\sigma)} \leqslant C_p \|f\|_{L^p(d\sigma)}.$$

Once again the work of A. P. Calderón [7] and Fabes, Jodeit, and Rivière [16] extends this theorem to C^1 domains. For Lipschitz domains, see Section 4.

2. HARMONIC MEASURE

After some opening remarks of a general nature, we will abandon the outline form of the first section and proceed with a more detailed, selfcontained exposition. The ideas of this section are motivated by the analysis given above of the Dirichlet problem for the ball in terms of the Poisson kernel.

Let D be a bounded, connected open set in $\mathbb{R}^n$. Let f be continuous and real-valued on ∂D. The maximum principle (1.18) and (1.22) imply that the harmonic function H_f given by the Perron process satisfies $|H_f(X)| \leqslant \sup |f|$. Therefore, the mapping $f \mapsto H_f(X)$ for any fixed $X \in D$ defines a continuous (positive) linear functional on $C^0(\partial D)$. By the Riesz representation theorem, there is a unique positive Borel measure ω^X on ∂D such that

$$H_f(X) = \int_{\partial D} f(Q) \, d\omega^X(Q).$$

In particular, $H_f(X) \equiv 1$ when $f \equiv 1$, and hence $\int_{\partial D} d\omega^X(Q) = 1$. In other words, ω^X has total mass one. It follows from Harnack's inequality (1.17) that ω^{X_1} and ω^{X_2} are mutually absolutely continuous. In fact, $d\omega^{X_1} = h \, d\omega^{X_2}$, where h is measurable $d\omega^{X_2}$ and $M^{-1} < h < M$, where M depends only on X_1, X_2, and D. Therefore, for any p, $1 \leqslant p \leqslant \infty$, $L^p(d\omega^{X_1}) = L^p(d\omega^{X_2})$, with comparable norms.

(2.1) THEOREM. *If f belongs to $L^1(d\omega^X)$ for some and hence all $X \in D$, then $u(X) = \int_{\partial D} f(Q) \, d\omega^X(Q)$ is harmonic in D and u is obtained from f by the Perron process.*

Fix $X_0 \in D$ and put $\omega = \omega^{X_0}$. We will call ω harmonic measure for D.

(2.2) THEOREM. *Let D be regular for the Dirichlet problem. A Borel set $E \subset \partial D$ has harmonic measure zero if and only if there is a positive harmonic function u in D such that $\lim_{X \to Q} u(X) = +\infty$ for every $Q \in E$.*

Now we will restrict our attention to Lipschitz domains. Our main goal in this section is to prove

(2.3) THEOREM. *Let D be a Lipschitz domain. If u is a positive harmonic function in D, then u has finite nontangential limits for almost every $(d\omega)$ $Q \in \partial D$.* (See 1.29 for the definition of nontangential limits.)

The mutual absolute continuity of harmonic measure (ω) and surface measure (σ) will be proved in Section 3.

The proof of (2.3) is based on several fundamental lemmas that will also be useful later. The proofs of these lemmas are deferred until Section 5.

For $Q \in \partial D$, a surface ball is defined by $\Delta(Q, r) = B(Q, r) \cap \partial D$. In all the statements that follow, D is a bounded Lipschitz domain.

(2.4) LEMMA. $\omega(\Delta(Q, 2r)) \leqslant C\omega(\Delta(Q, r))$, *for a constant C independent of $Q \in \partial D$ and $r > 0$.*

For $f \in L^1(\partial D, d\omega)$, let

$$\mathfrak{M}_\omega f(Q) = \sup\left\{ \frac{1}{\omega(\Delta)} \int_\Delta |f| \, d\omega : Q \in \Delta, \Delta \text{ is a surface ball} \right\}.$$

The importance of Lemma (2.4) is that it and the Vitali covering lemma [44] imply the generalized Hardy-Littlewood maximal theorem:

(2.5) THEOREM. *There exists C such that*

$$\omega\{Q \in \partial D: \mathfrak{M}_\omega f(Q) > \lambda\} \leqslant \frac{C}{\lambda} \int_{\partial D} |f| \, d\omega$$

for $\lambda > 0$, $f \in L^1(\partial D, d\omega)$. Moreover, if $f \in L^p(\partial D, d\omega)$,

$$\|\mathfrak{M}_\omega f\|_{L^p(d\omega)} \leqslant C_p \|f\|_{L^p(d\omega)}, \quad \text{for } 1 < p \leqslant \infty.$$

Recall that for u defined in D,

$$N_\alpha(u)(Q) = \sup\{|u(X)| : X \in \Gamma_\alpha(Q)\}, \qquad Q \in \partial D.$$

(2.6) LEMMA. *If $f \in L^1(d\omega)$ and $u(X) = \int_{\partial D} f \, d\omega^X$, then $N_\alpha(u)(Q) \leqslant C_\alpha \mathfrak{M}_\omega f(Q)$, where C_α depends only on D and α.*

For any Lipschitz domain D, there exist numbers M and $r_0 > 0$ such that if $r < r_0$ and $Q \in \partial D$, there is a point $A = A_r(Q) \in D$ such that $M^{-1}r < |A - Q| < Mr$ and $\operatorname{dist}(A, \partial D) > M^{-1}r$. This is easy to see from the existence of interior cones like the exterior cones considered in Section 1. A tool in proving Lemma (2.6) that will also be used in the next section is

(2.7) LEMMA. *Assume that $r < r_0$ and E is a Borel subset of $\Delta(Q, r/2)$. Then*

$$\omega^{A_r(Q)}(E) \simeq \omega(E)/\omega(\Delta(Q, r)).$$

(Here and elsewhere, the notation $R \simeq S$ means that R/S is bounded between two positive constants depending only on the domain D.)

We can now prove a preliminary version of (2.3).

(2.3') THEOREM. *Suppose that $1 \leqslant p \leqslant \infty$, and $f \in L^p(\partial D, d\omega)$. Let $u(X) = \int_{\partial D} f \, d\omega^X$. Then u has nontangential limit f a.e. $d\omega$ on ∂D, and*

(a) $\quad \|N_\alpha(u)\|_{L^p(d\omega)} \leqslant C_{p,\alpha}\|f\|_{L^p(d\omega)}$ for $1 < p \leqslant \infty$;

(b) $\quad \omega\{Q \in \partial D : N_\alpha(u) > \lambda\} \leqslant \dfrac{C_\alpha}{\lambda}\|f\|_{L^1(d\omega)}$.

Proof. Inequalities (a) and (b) follow from (2.5) and (2.6). Since $L^p(d\omega) \subset L^1(d\omega)$ for $p > 1$, it is enough to prove the existence of nontangential limits for $f \in L^1(d\omega)$. We will deduce this from (b) by a well-known argument (see [44] for example).

Let $f \in L^1(d\omega)$ and fix $\alpha > 0$. Set $u(X) = \int_{\partial D} f \, d\omega^X$,

$$\Lambda(f)(Q) = \left| \limsup_{\substack{X \to Q \\ X \in \Gamma_\alpha(Q)}} u(X) - \liminf_{\substack{X \to Q \\ X \in \Gamma_\alpha(Q)}} u(X) \right|.$$

We will show that $\Lambda(f) = 0$ a.e. $d\omega$. If g is continuous, then by Proposition (1.26), $v(X) = \int_{\partial D} g \, d\omega^X$ is continuous in $\overline{D}$. Hence $\Lambda(g) = 0$ everywhere. Let $\varepsilon > 0$. Choose a continuous function g such that $\|f - g\|_{L^1(d\omega)} \leqslant \varepsilon$. Then

$$\Lambda(f)(Q) \leqslant \Lambda(g)(Q) + \Lambda(f - g)(Q)$$
$$= \Lambda(f - g)(Q) \leqslant 2N_\alpha(u - v)(Q).$$

Therefore,

$$\omega\{Q \in \partial D. \ \Lambda(f)(Q) > \lambda\} \leqslant \omega\{Q \in \partial D: 2N_\alpha(u - v)(Q) > \lambda\}$$
$$\leqslant \frac{2C_\alpha}{\lambda} \|f - g\|_{L^1(d\omega)} \leqslant \frac{2C_\alpha}{\lambda} \varepsilon.$$

Since ε is arbitrary, we conclude that $\Lambda(f)(Q) = 0$ a.e. $Q, d\omega$. Thus u has nontangential limit almost everywhere $d\omega$.

Denote $h(Q) = \lim_{X \to Q} u(X)$, $X \in \Gamma_\alpha(Q)$. It remains to show that $h(Q) = f(Q)$ a.e. $Q, d\omega$. Let

$$h_1(Q) = \lim_{X \to Q} [u(X) - v(X)], \ X \in \Gamma_\alpha(Q).$$

Then $h - f = h_1 + g - f$. Hence,

$$\omega\{Q \in \partial D: |h(Q) - f(Q)| > \lambda\} \leqslant \omega\left\{Q \in \partial D: |h_1(Q)| > \frac{\lambda}{2}\right\}$$

$$+ \omega\{Q \in \partial D: |g(Q) - f(Q)| > \lambda/2\}$$

$$\leqslant \omega\{Q \in \partial D: N_\alpha(u - v)(Q) > \lambda/2\}$$

$$+ \omega\{Q \in \partial D: |g(Q) - f(Q)| > \lambda/2\} \leqslant \frac{2}{\lambda}(C_\alpha + 1)\|f - g\|_{L^1(d\omega)}$$

$$\leqslant \frac{2}{\lambda}(C_\alpha + 1)\varepsilon.$$

Since ε is arbitrary, we see that $f(Q) = h(Q)$ a.e. $Q, d\omega$, as desired.

(2.8) *Definition.* Let $K(X, Q) = d\omega^X/d\omega(Q)$ be the Radon-Nykodim derivative of ω^X with respect to $\omega(= \omega^{X_0})$.

By our introductory remarks, $K(X, Q)$ exists a.e. $Q \, d\omega$ and is a bounded function of Q. $K(X, Q)$ is known as the *kernel function* of D.

Example. Let D be the unit ball, and let X_0 be the center of the ball. We already noted that the mean value theorem says that $d\omega^{X_0} = \omega_n^{-1} \, d\sigma$, where $d\sigma$ is surface measure on the boundary of the ball and ω_n is the total measure of the boundary. Thus

$$d\omega^X(Q) = P(X, Q) \, d\sigma(Q) = \omega_n P(X, Q) \, d\omega^{X_0}(Q),$$

and K is just a constant multiple of the Poisson kernel $K(X, Q) = \omega_n P(X, Q)$.

(2.9) LEMMA. *For every* $Q \in \partial D$, $\lim \dfrac{\omega^X(\Delta)}{\omega(\Delta)}$ *exists as the surface balls* Δ *shrink to* Q. *The limit defines* $K(X, Q)$ *for every* Q, *and* $K(X, Q)$ *is continuous as a function of* Q. *In fact,* $K(X, Q)$ *is Hölder continuous as a function of* Q:

$$|K(X, Q) - K(X, Q')| \leq C_X |Q - Q'|^\alpha,$$

where α *depends only on* D.

Once again, the proof of this lemma is deferred to Section 5. At that time we will also describe the kernel function in another way. $K(X, Q)$ is the unique positive harmonic function of X in D satisfying $K(X_0, Q) = 1$ and vanishing continuously as $X \to Q'$ for all $Q' \in \partial D$, $Q' \neq Q$. Let μ be a finite Borel measure on ∂D. From this remark and Lemma (2.9) we see that

$$u(X) = \int_{\partial D} K(X, Q) \, d\mu(Q)$$

is a well-defined harmonic function in D. Denote

$$\mathfrak{M}_\omega(d\mu)(Q) = \sup\left\{\frac{1}{\omega(\Delta)} \int_\Delta |d\mu| : Q \in \Delta, \Delta \text{ is a surface ball on } \partial D\right\}.$$

When we prove (2.6) we will also show $N_\alpha(u)(Q) \leqslant C_\alpha \mathfrak{M}_\omega(d\mu)$. Moreover, the same proof that gave Theorem (2.5) shows

$$\omega\{Q \in \partial D: \mathfrak{M}_\omega(d\mu) > \lambda\} \leqslant \frac{C}{\lambda} \int_{\partial D} |d\mu|.$$

It is easy to extend the method of proof of (2.3′) to show that if μ is singular with respect to ω, then $u(X)$ converges nontangentially to 0 a.e. $d\omega$. Thus, we have established

(2.3″) THEOREM. *Let μ be a finite Borel measure on ∂D. Let $u(X) = \int_{\partial D} K(X, Q) \, d\mu(Q)$. Then $u(X)$ converges nontangentially a.e. $d\omega$ to a limit $f \in L^1(d\omega)$, where $d\mu = f \, d\omega + \mu_s$, and μ_s is singular with respect to ω.*

To prove Theorem (2.3), all that remains is to show that every positive harmonic function in D can be written in the form $\int_{\partial D} K(X, Q) \, d\mu(Q)$ for some positive Borel measure μ. While this is true [25], it is somewhat easier to prove it for starlike Lipschitz domains ((2.10) below) and to complete the proof by realizing any Lipschitz domain as a union of starlike Lipschitz domains.

(2.10) *Definition.* Let (r, θ) be polar coordinates for $\mathbb{R}^n$, $0 \leqslant r < \infty$, $\theta \in S^{n-1}$, the unit sphere in $\mathbb{R}^n$. A domain D in $\mathbb{R}^n$ is a *starlike Lipschitz domain* (with respect to the origin) if there exists $\phi: S^{n-1} \to \mathbb{R}$, ϕ is strictly positive, and $|\phi(\theta) - \phi(\theta')| \leqslant M|\theta - \theta'|$ for all $\theta, \theta' \in S^{n-1}$, so that $D = \{(r, \theta): 0 \leqslant r < \phi(\theta)\}$.

Consider an arbitrary Lipschitz domain D and choose coordinates as in Definition (1.28) and a ball B so that

$$B \cap D = \{X = (x', x_n): x_n > \phi(x')\} \cap B$$

and $\phi: \mathbb{R}^{n-1} \to \mathbb{R}$ is a Lipschitz function with Lipschitz constant M. For appropriate $t > 0$, and $a > 0$, $b > 0$ depending only on M, the domain $D \cap U$ is a starlike Lipschitz domain with respect to X_0 where $X_0 = (0, bt)$,

$$U = \{(x', x_n): |x'| < t, |x_n| < at\}.$$

Therefore, for the study of the boundary behavior of a function u on D it suffices to consider $D \cap U$, a starlike Lipschitz domain. Furthermore, one can show [24] that the restriction of harmonic measure for D and $D \cap U$ to a closed subset of $U \cap \partial D$ are mutually absolutely continuous. Hence, to establish Theorem (2.3) it is enough to prove it for starlike Lipschitz domains.

Now denote by D a Lipschitz domain that is starlike with respect to the origin. Let u be harmonic and positive in D. Denote

$$u_r(X) = u(rX), \qquad 0 \leqslant r < 1.$$

Since u_r is continuous in $\overline{D}$,

$$u_r(X) = \int_{\partial D} u(rQ) \, d\omega^X(Q) = \int_{\partial D} K(X, Q) u(rQ) \, d\omega(Q).$$

Let $X_0 = 0$. ($\omega = \omega^{X_0} = \omega^0$.) Then $u(0) = u_r(0) = \int_{\partial D} u(rQ) \, d\omega(Q)$. The positive measures $u(rQ) \, d\omega(Q)$ have uniformly bounded (in fact, constant) total variation. Therefore, there is a sequence $r_n \to 1$ such that $u(r_n Q) \, d\omega(Q) \to d\mu(Q)$ weakly, for some positive measure μ. Since $K(X, Q)$ is continuous in Q,

$$u(X) = \lim_{n \to \infty} u_{r_n}(X) = \int_{\partial D} K(X, Q) \, d\mu(Q).$$

Theorem (2.3) now follows from (2.3″).

3.   THE DIRICHLET PROBLEM IN LIPSCHITZ DOMAINS

In this section we prove two theorems of B. E. J. Dahlberg.

(3.1) THEOREM.   *Let D be a Lipschitz domain. Then*

(a)   *Harmonic measure ω is absolutely continuous with respect to surface measure σ on ∂D.*

(b)   Let $k(Q) = \dfrac{d\omega}{d\sigma}(Q)$. Then $k \in L^2(d\sigma)$. Moreover, for every surface ball $\Delta \subset \partial D$,

$$\left(\frac{1}{\sigma(\Delta)} \int_\Delta k^2 \, d\sigma \right)^{1/2} \leqslant \frac{C}{\sigma(\Delta)} \int_\Delta k \, d\sigma.$$

   The constant C depends only on D.

(c)   Surface measure is absolutely continuous with respect to harmonic measure.

(3.2) COROLLARY.   Let D be a Lipschitz domain. If $p \geqslant 2$ and $f \in L^p(\partial D, d\sigma)$, then there is a unique harmonic function u in D such that u converges nontangentially a.e. $d\sigma$ to f and

$$\| N_\alpha(u) \|_{L^p(d\sigma)} \leqslant C_{p,\alpha} \| f \|_{L^p(d\sigma)}.$$

For simplicity we will only prove these theorems for starlike Lipschitz domains (see (2.10)). The proof can be modified slightly to handle general Lipschitz domains [27]. We will also assume $n \geqslant 3$. (See (6.8) for remarks on the easier case $n = 2$.)

The first lemma is the key to the entire development.

(3.3) LEMMA.   Let D be a bounded C^∞ domain in $\mathbb{R}^n$ containing the origin. Let ω denote harmonic measure for D at 0 and $k = d\omega/d\sigma$. Then

$$\frac{1}{\omega_n} \int_{\partial D} \frac{k(Q)}{|Q|^{n-2}} \, d\sigma(Q) = \int_{\partial D} k^2(Q) \langle Q, N_Q \rangle \, d\sigma(Q),$$

where N_Q is the outer unit normal and $\langle \, , \, \rangle$ is the usual inner product.

Proof: By (1.13), the Green function for D with pole at 0 is given by

$$G(X) = \frac{-1}{(n-2)\omega_n} |X|^{n-2} - u(X) = F(X) - u(X),$$

where $u(X)$ is harmonic in D and G is zero on ∂D. The method of layer potentials implies that $G \in C^\infty(\overline{D} \setminus \{0\})$. Thus the formal reasoning behind (1.14) can be justified and $k(Q) = \dfrac{\partial G}{\partial N_Q}(Q)$. Denote

$$\alpha(Q) = Q - \langle Q, N_Q \rangle N_Q.$$

Because $\langle \alpha(Q), N_Q \rangle = 0$, and G is zero on ∂D, $\langle \alpha(Q), \nabla G(Q) \rangle = 0$ for $Q \in \partial D$. Hence,

$$\int_{\partial D} \langle \alpha(Q), \nabla G(Q) \rangle k(Q) \, d\sigma(Q) = 0.$$

In other words,

$$\int_{\partial D} k(Q) \langle Q, \nabla G(Q) \rangle \, d\sigma(Q) = \int_{\partial D} k^2(Q) \langle Q, N_Q \rangle \, d\sigma(Q).$$

Let $v(X) = \langle X, \nabla u(X) \rangle$. A direct computation shows that v is harmonic. Moreover, $v(0) = 0$. Therefore,

$$\int_{\partial D} \langle Q, \nabla u(Q) \rangle k(Q) \, d\sigma(Q) = v(0) = 0,$$

and the lemma follows.

Lemma (3.3) will be used to establish an *a priori* inequality. To pass from C^∞ domains to Lipschitz domains we need two lemmas.

(3.4) LEMMA. *Let* $\phi : S^{n-1} \to \mathbb{R}$ *satisfy* $\phi(\theta) \geq \beta > 0$ *and* $|\phi(\theta) - \phi(\theta')| \leq M |\theta - \theta'|$ *for all* $\theta, \theta' \in S^{n-1}$. *Then there exists a sequence of functions* $\phi_m \in C^\infty(S^{n-1})$ *with* $\phi_m(\theta) \geq \beta/2$, $\phi_m(\theta)$ *increases to* $\phi(\theta)$ *uniformly,* $\nabla_\theta \phi_m(\theta) \to \nabla_\theta \phi(\theta)$ *almost everywhere* $d\theta$, *and* $\sup_\theta |\nabla_\theta \phi_m(\theta)| \leq 2M$.

Proof. For Lipschitz functions $\phi : \mathbb{R}^{n-1} \to \mathbb{R}$, the corresponding lemma is proved by taking the convolution $\phi * \psi_m$ of ϕ with an approximate identity $\psi_m(x) = m^{-n} \psi(x/m)$, where $\psi \in C_0^\infty(\mathbb{R}^{n-1})$, $\int \psi(x) \, dx = 1$. A slight modification is made to insure that the sequence is increasing. The result for the sphere is reduced to the case of $\mathbb{R}^{n-1}$ by using local coordinates on S^{n-1} and a partition of unity.

(3.5) LEMMA. *Assume that ϕ and ϕ_m are as in (3.4) $D = \{(r, \theta): 0 \leqslant r < \phi(\theta)\}$ and $D_m = \{(r, \theta): 0 \leqslant r < \phi_m(\theta)\}$. Let ω (resp. ω^m) denote harmonic measure for D (resp. D_m) at 0. Let $\tilde{\omega}_m$ be the measure on ∂D defined by $\int_{\partial D} f \, d\tilde{\omega}_m = \int_{\partial D_m} \tilde{f}_m \, d\omega^m$, where $\tilde{f}_m(\phi_m(\theta), \theta) = f(\phi(\theta), \theta)$. (Recall that $\partial D_m = \{(\phi_m(\theta), \theta): \theta \in S^{n-1}\}$ in polar coordinates.) Then, $\tilde{\omega}_m$ converges weakly to ω as $m \to \infty$.*

Proof. Let $f \in C^0(\partial D)$ and $u(X) = \int_{\partial D} f \, d\omega^X$. Let $f_m \in C^0(\partial D_m)$ be defined by $f_m = u|_{\partial D_m}$. Since D is regular for the Dirichlet problem (1.26), $\lim_{m \to \infty} \sup_{\partial D_m} |f_m - \tilde{f}_m| = 0$. Therefore, $\lim_{m \to \infty} \int_{\partial D_m} (\tilde{f}_m - f_m) \, d\omega^m = 0$. Moreover, for all m, $\int_{\partial D_m} f_m \, d\omega^m = \int_{\partial D} f \, d\omega$, and our result follows.

Proof of (3.1). Let D_m be the C^∞ domains of (3.5), and let $k_m = d\omega^m/d\sigma_m$. ($\omega^m$ and σ_m are harmonic measure and surface measure on ∂D_m, respectively.) A simple calculation shows that if $Q \in \partial D_m$ and N_Q denotes the outer unit normal to ∂D_m at Q, then

$$\langle Q, N_Q \rangle = \frac{\phi_m(\theta)}{\left(1 + |\nabla_\theta \phi_m(\theta)|^2\right)^{1/2}},$$

where $Q = (\phi_m(\theta), \theta)$. Therefore,

$$\langle Q, N_Q \rangle \geqslant \frac{\beta}{2} \frac{1}{(1 + 4M^2)^{1/2}}$$

where β and M are as in (3.4). An application of (3.3) shows that

$$\frac{\beta}{2} \frac{1}{(1 + 4M^2)^{1/2}} \int_{\partial D_m} k_m^2 \, d\sigma_m \leqslant \frac{1}{\omega_n} \int_{\partial D_m} \frac{k_m(Q)}{|Q|^{n-2}} \, d\sigma_m$$

$$\leqslant \frac{1}{\omega_n (\beta/2)^{n-2}} \int_{\partial D_m} k_m(Q) \, d\sigma_m(Q)$$

$$= \frac{1}{\omega_n (\beta/2)^{n-2}}.$$

Thus

$$\int_{\partial D_m} k_m^2 \, d\sigma_m \leqslant C_M / \beta^{n-1}, \text{ where } C_M \text{ depends only on } M. \quad (3.6)$$

Consider $\tilde{\omega}_m$ defined in (3.5). Let $f \in C^0(\partial D)$.

$$\left| \int_{\partial D} f \, d\tilde{\omega}_m \right| = \left| \int_{\partial D_m} \tilde{f}_m \, d\omega_m \right| = \left| \int_{\partial D_m} \tilde{f}_m k_m \, d\sigma_m \right|$$

$$\leqslant \left(\int_{\partial D_m} |\tilde{f}_m|^2 \, d\sigma_m \right)^{1/2} \left(\int_{\partial D_m} k_m^2 \, d\sigma_m \right)^{1/2}$$

$$\leqslant C' \left(\int_{\partial D_m} |\tilde{f}_m|^2 \, d\sigma_m \right)^{1/2} \leqslant C'' \|f\|_{L^2(\partial D, d\sigma)}$$

for a constant C'' depending only on D. Therefore, there exist functions $\tilde{k}_m \in L^2(\partial D, d\sigma)$ such that $d\tilde{\omega}_m = \tilde{k}_m \, d\sigma$ and $\|k_m\|_{L^2(\partial D, d\sigma)} \leqslant C''$, independent of m. Hence, there is a subsequence m_j and a function $k \in L^2(\partial D, d\sigma)$ with $\tilde{k}_{m_j} \to k$ weakly in $L^2(d\sigma)$. The measures $\tilde{\omega}_{m_j}$ converge weakly to $k \, d\sigma$, and hence by (3.5), $d\omega = k \, d\sigma$. Therefore, (a) is established.

For surface balls of large radius (b) is contained in (a), so we may as well assume that the radius r is small, i.e., $r \ll \beta$ and $r \ll M^{-1}$. We will define a domain $\tilde{D} \supset D$ as follows:

$$Q \in \partial D, A_1 = (1 - 4Mr)Q, A_2 = (1 - 2Mr)Q.$$

$L = \{A_2 + s(Q - A_2): 0 \leqslant s < \infty\}$, the ray from A_2 in the direction of Q.

$$\mathcal{C} = \left\{ X: \text{dist}(X, L) \leqslant \left(\sin \frac{1}{2M} \right) |X - A_2| \right\}$$

is a cone with vertex at A_2 and axis L. A surface ball Δ about Q of radius (essentially) r is given by $\Delta = \mathcal{C} \cap \partial D$. Let $R = \sup_\theta |\phi(\theta)|$.

$$\tilde{D} = (\mathcal{C} \cap D) \cup (B(A_1, 2R) \backslash \mathcal{C}).$$

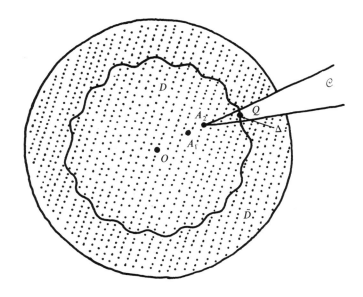

The reader can check that

(a) $\tilde{D}$ is starlike with respect to A_1 with Lipschitz constant M' comparable to M;
(b) the lower bound β' corresponding to β in (3.4) and in the proof of (3.1) (a) satisfies

$$\beta' = \text{dist}\left(A_1, \partial\tilde{D}\right) \geqslant r.$$

Denote by $\tilde{\omega}^A$ harmonic measure for $\tilde{D}$ at A and $\tilde{k}_A = d\tilde{\omega}^A/d\tilde{\sigma}$, where $\tilde{\sigma}$ is the surface measure of $\partial\tilde{D}$. By (3.6) we see that $\int_{\partial\tilde{D}} \tilde{k}_A^2 \, d\tilde{\sigma} \leqslant C_M/r^{n-1}$. Observe that $\partial\tilde{D} \cap \partial D = \Delta$, $\tilde{\sigma} = \sigma$ on Δ. Since $\tilde{D} \supset D$, the maximum principle implies that $d\omega^A/d\sigma \leqslant \tilde{k}_A$ a.e. $d\sigma$ on Δ. Let $k_A = d\omega^A/d\sigma$. Then

$$\int_\Delta k_A^2 \, d\sigma \leqslant C_M/r^{n-1} \simeq C_M/\sigma(\Delta).$$

Finally Lemma (2.7) shows that $k \leqslant C_M k_A \omega(\Delta)$ a.e. $d\sigma$ on Δ.

Therefore

$$\int_\Delta k^2 \, d\sigma \leqslant C_M \frac{\omega(\Delta)^2}{\sigma(\Delta)} = C_M \frac{\left(\int_\Delta k \, d\sigma\right)^2}{\sigma(\Delta)},$$

and (b) follows. Part (c) is a consequence of (b) alone. (See [22], [38], [10], [12].)

Condition (b) has a remarkable consequence proved by Gehring [22] and Muckenhoupt [38], namely an improvement of itself: $k \in L^{2+\varepsilon}(d\sigma)$ and

(b') $$\left(\frac{1}{\sigma(\Delta)} \int_\Delta k^{2+\varepsilon} d\sigma\right)^{1/(2+\varepsilon)} \leqslant \frac{C'}{\sigma(\Delta)} \int_\Delta k \, d\sigma,$$

for some $\varepsilon > 0$ and C' depending only on C in (b).

Remark. By examining angles in $\mathbb{R}^2$ (where one can compute k using conformal mapping) one can see that while $k \in L^{2+\varepsilon}(d\sigma)$, the value of ε tends to zero as the Lipschitz constant tends to infinity. Thus the estimate (b') is in the nature of best possible. On the other hand, for C^1 domains k satisfies (b') for arbitrarily large ε ([13], [18]).

Proof of Corollary (3.2).   Let $p \geqslant 2$, $f \in L^p(\partial D, d\sigma)$. Since $k \in L^2(d\sigma)$, Schwarz' inequality implies $f \in L^1(d\omega)$. Therefore, by (2.3') $u(X) = \int_{\partial D} f \, d\omega^X$ has nontangential limit equal to f a.e. $d\omega$ and hence by (3.1) (c) a.e. $d\sigma$. By (2.6), $N_\alpha(u)(Q) \leqslant C_\alpha \mathfrak{M}_\omega(f)(Q)$, and therefore the bound on $N_\alpha(u)$ follows from the inequality

$$\|\mathfrak{M}_\omega(f)\|_{L^p(d\sigma)} \leqslant C_p \|f\|_{L^p(d\sigma)}, \qquad p \geqslant 2. \qquad (3.7)$$

We will prove (3.7) for $p > 2$ first. By Schwarz' inequality and (3.1) (b),

$$\frac{1}{\omega(\Delta)} \int_\Delta |f| \, d\omega \leqslant \frac{1}{\omega(\Delta)} \left(\int_\Delta |f|^2 d\sigma\right)^{1/2} \left(\int_\Delta k^2 d\sigma\right)^{1/2}$$

$$\leqslant C \left(\frac{1}{\sigma(\Delta)} \int_\Delta |f|^2 d\sigma\right)^{1/2}.$$

Therefore,

$$\mathfrak{M}_\omega(f)(Q) \leqslant C\mathfrak{M}_\sigma\big(|f|^2\big)(Q)^{1/2}.$$

Since σ satisfies the doubling condition, $\sigma(\Delta(Q,2r)) \leqslant C\sigma(\Delta(Q,r))$ we have, as in (2.5),

$$\|\mathfrak{M}_\sigma(g)\|_{L^q(d\sigma)} \leqslant C_q\|g\|_{L^q(d\sigma)} \quad \text{for } 1 < q < \infty.$$

If $f \in L^p(d\sigma)$, $p > 2$, then

$$\int_{\partial D} |\mathfrak{M}_\omega f|^p \, d\sigma \leqslant C\int_{\partial D} \mathfrak{M}_\sigma\big(|f|^2\big)^{p/2} \, d\sigma$$

$$\leqslant C'\int_{\partial D} \big(|f|^2\big)^{p/2} \, d\sigma = C'\int_{\partial D} |f|^p \, d\sigma.$$

For $p = 2$, this argument does not apply because $\mathfrak{M}_\sigma$ does not map $L^1(d\sigma)$ to $L^1(d\sigma)$. However, if we apply (b') instead of (b) and Hölder's inequality rather than Schwarz' inequality, we find that

$$\mathfrak{M}_\omega(f)(Q) \leqslant C\mathfrak{M}_\sigma\big(|f|^{2-\varepsilon}\big)(Q)^{1/(2-\varepsilon)}$$

for some $\varepsilon > 0$, and the estimate for $p = 2$ follows. The uniqueness follows from Lebesgue's dominated convergence theorem and a procedure like the one used in (2.3) at the very end of Section 2.

Remark. Once again by looking at angles in $\mathbb{R}^2$, one can see that the exponent 2 is sharp in that given any $p < 2$ there is a Lipschitz domain and a function $f \in L^p(d\sigma)$ such that the conclusion of (3.2) is false. For C^1 domains, it is easy to see, using the fact that (b') is valid for arbitrarily large $\varepsilon > 0$, that (3.2) holds for $f \in L^p(d\sigma)$ for any $p > 1$. (See [13].) As mentioned before this can also be obtained by means of layer potentials [16].

4. THE NEUMANN PROBLEM IN LIPSCHITZ DOMAINS

Once again for simplicity we will only discuss starlike Lipschitz domains (2.10).

(4.1) THEOREM. *Let D be a starlike Lipschitz domain. Given $f \in L^2(\partial D, d\sigma)$ with $\int_{\partial D} f \, d\sigma = 0$, there exists a harmonic function u, unique up to an additive constant such that*

$$\|N_\alpha(\nabla u)\|_{L^2(d\sigma)} \leqslant C\|f\|_{L^2(d\sigma)}.$$

∇u *has nontangential limits $\nabla u(Q)$ a.e. $d\sigma$, and $\langle N_Q, \nabla u(Q)\rangle = f(Q)$ a.e. $d\sigma$.*

Unlike the solution to the Dirichlet problem considered in Section 3, the solution to the Neumann problem can already be shown to exist in a weaker sense than that of (4.1) by means of a general Hilbert space procedure. We will first describe this so-called weak solution. For this purpose we introduce the Sobolev spaces $H_s(D)$.

(4.2) *Definition.* Let $0 \leqslant s < \infty$. $H_s(D) = \{f \in L^2(D): f = h|_D$ where $h \in L^2(\mathbb{R}^n)$ and $\int_{\mathbb{R}^n} |\hat{h}(\xi)|^2 (1 + |\xi|^2)^s \, d\xi < \infty\}$. ($\hat{h}$ is the Fourier transform of h.) These spaces can also be defined intrinsically. For example, $H_1(D)$ is the closure of $C^\infty(\overline{D})$ under the norm

$$\|h\| = \left(\int_D h^2\right)^{1/2} + \left(\int_D |\nabla h|^2\right)^{1/2}. \qquad (4.3)$$

Another useful remark is that if $h \in H_1(D)$, then h has a well-defined restriction to ∂D and

$$\left(\int_{\partial D} h^2 \, d\sigma\right)^{1/2} \leqslant C\|h\|, \qquad (4.4)$$

where the constant C depends only on the Lipschitz constant of D. This can be proved by observing that if $h \in H_1(D)$ and h vanishes in a neighborhood of 0, then $h(r/\phi(\theta), \theta)$ is in H_1 of the unit ball, with suitable control of its norm. The corresponding statement for the ball is elementary ([21, p. 113]).

(4.5) *Definition.* A *weak solution* of the Neumann problem for boundary data f is a function $u \in H_1(D)$ such that, for all $h \in$

$H_1(D)$,

$$\int_D \langle \nabla u(X), \nabla h(X) \rangle \, dX = \int_{\partial D} f(Q) h(Q) \, d\sigma(Q).$$

It is easy to see from Green's formula (1.5) that if D is smooth, $u \in C^1(\overline{D})$, $\Delta u = 0$ in D and $\partial u / \partial N = f$, then u is a weak solution to the Neumann problem (4.5).

(4.6) THEOREM. *Let D be a starlike Lipschitz domain. Given $f \in L^2(d\sigma)$ with $\int_{\partial D} f \, d\sigma = 0$, there exists $u \in H_1(D)$ unique up to an additive constant that is a weak solution to the Neumann problem with boundary data f.*

Proof. Denote by $\dot{H}_1(D)$ the space $H_1(D)$ modulo constants. An element of $\dot{H}_1(D)$ will be identified with any of its representatives. The norm is $\|h\|_{\dot{H}_1} = \inf\{\|h + c\| : c \in \mathbb{R}\}$ ($\|\ \|$ is given by (4.3)). Consider the bilinear form on $\dot{H}_1(D) \times \dot{H}_1(D)$, $b(u, h) = \int_D \langle \nabla u, \nabla h \rangle$. The main point is that $b(h, h)^{1/2}$ is equivalent to the $\dot{H}_1(D)$ norm $\|h\|_{\dot{H}_1}$. To prove this all we need to do is to find a representative of h such that $\int_D h^2 \leqslant c \int_D |\nabla h|^2$. It suffices to do this on a dense subclass $h \in C^\infty(\overline{D})$. In that case we can specify h by $h(0) = 0$ and write

$$h(X) = \int_0^1 \langle X, \nabla h(sX) \rangle \, ds. \tag{4.7}$$

Using (4.7), it is an easy matter to show that $\int_D h^2 \leqslant C \int_D |\nabla h|^2$ with a constant depending on D but not on h.

We are now in a position to apply the lemma of Lax and Milgram (a general abstract lemma that applies to any Hilbert space).

(4.8) LEMMA. *If $b(h, h)^{1/2}$ is equivalent to the $\dot{H}_1(D)$ norm $\|h\|_{\dot{H}_1}$, and λ is a continuous linear functional on $\dot{H}_1$, then there is a unique $u \in \dot{H}_1(D)$ such that*

$$b(u, h) = \lambda(h).$$

By (4.4), if $f \in L^2(d\sigma)$, then $\lambda(h) = \int_{\partial D} fh \, d\sigma$ is a continuous linear functional on $H_1(D)$. If in addition $\int f \, d\sigma = 0$, then λ annihilates constants, so that it is a functional on $\dot{H}_1(D)$. By the Lax-Milgram lemma, there is a unique $u \in \dot{H}_1(D)$ such that for all $h \in H_1(D)$,

$$\int_D \langle \nabla u(X), \nabla h(X) \rangle \, dX = \int_{\partial D} f(Q) h(Q) \, d\sigma(Q).$$

Remark 1.   If u is a weak solution of the Neumann problem, then u is harmonic in D. In fact, $\int_D \langle \nabla u, \nabla \phi \rangle = 0$ for all $\phi \in C_0^\infty(D)$ implies that $\Delta u = 0$ in D in the sense of distributions and hence ([21]) $u \in C^\infty(D)$ and $\Delta u = 0$.

Remark 2.   Let D_m be the domains of (3.5). If u is a weak solution to the Neumann problem with boundary data f, then

$$\lim_{m \to \infty} \int_{\partial D_m} \frac{\partial u}{\partial N_m} h \, d\sigma_m = \int_{\partial D} fh \, d\sigma$$

for every $h \in H_1(D)$. This is true because $\nabla u \in L^2(D)$ and $\nabla h \in L^2(D)$ implies $\lim_{m \to \infty} \int_{D \setminus D_m} \langle \nabla u, \nabla h \rangle = 0$ and

$$\int_{D_m} \langle \nabla u, \nabla h \rangle = \int_{\partial D_m} \frac{\partial u}{\partial N_m} h \, d\sigma_m$$

by Green's formula (1.5).

Remark 3.   Let u be the weak solution as in 4.6. Then $\|u\|_{\dot{H}_1(D)}$ $\leqslant C \|f\|_{L^2(d\sigma)}$, where C depends only on the Lipschitz constant of D.

We can now proceed with the proof of Theorem (4.1). The main idea is contained in the following two lemmas.

(4.9) LEMMA. *Let D be a C^∞ domain in $\mathbb{R}^n$ and let u be harmonic in D, $u \in C^\infty(\bar{D})$. Then*

$$\int_{\partial D} \left\{ \left[|\nabla_t u|^2 - \left(\frac{\partial u}{\partial N} \right)^2 \right] \langle Q, N_Q \rangle - 2 \langle \alpha(Q), \nabla u \rangle \frac{\partial u}{\partial N} \right.$$

$$\left. - (n-2) u \frac{\partial u}{\partial N} \right\} \, d\sigma(Q) = 0,$$

with the notations $\alpha(Q) = Q - \langle Q, N_Q \rangle N_Q$, ∇_t is the tangential component of ∇ at ∂D, i.e., $\nabla_t = (\langle T_1, \nabla \rangle, \ldots, \langle T_{n-1}, \nabla \rangle)$ where $N_Q, T_1, \ldots, T_{n-1}$ form an orthonormal basis at $Q \in \partial D$.

Proof. A computation shows that

$$\mathrm{div}\{ |\nabla u|^2 Q - 2 \langle Q, \nabla u \rangle \nabla u - (n-2) u \nabla u \}$$

$$= -2 \langle Q, \nabla u \rangle \Delta u - (n-2) u \Delta u = 0.$$

By the divergence theorem (1.4),

$$\int_{\partial D} \left\{ |\nabla u|^2 \langle Q, N_Q \rangle - 2 \langle Q, \nabla u \rangle \frac{\partial u}{\partial N} - (n-2) u \frac{\partial u}{\partial N} \right\} \, d\sigma(Q) = 0.$$

Since $|\nabla u|^2 = |\nabla_t u|^2 + (\partial u / \partial N)^2$ and

$$\langle Q, \nabla u \rangle = \langle \alpha(Q), \nabla u \rangle + \langle N_Q, \nabla_u \rangle \langle Q, N_Q \rangle$$

$$= \langle \alpha(Q), \nabla u \rangle + \langle Q, N_Q \rangle \frac{\partial u}{\partial N},$$

we have

$$\int_{\partial D} \left\{ |\nabla_t u|^2 \langle Q, N_Q \rangle + \left(\frac{\partial u}{\partial N} \right)^2 \langle Q, N_Q \rangle - 2 \left(\frac{\partial u}{\partial N} \right)^2 \langle Q, N_Q \rangle \right.$$

$$\left. - 2 \langle \alpha(Q), \nabla u \rangle \frac{\partial u}{\partial N} - (n-2) u \frac{\partial u}{\partial N} \right\} \, d\sigma(Q) = 0.$$

This is the desired formula.

We will need a slight variant of (3.4). The functions ψ_j we want satisfy the same properties as ϕ_m in (3.4) except that ψ_j decreases to ϕ. The resulting C^∞ domains will be called Ω_j. Obviously Ω_j decreases to D.

(4.10) LEMMA. *Let Ω be one of the Ω_j above and assume $f \in C^\infty(\mathbb{R}^n)$, $\int_{\partial\Omega} f \, d\sigma = 0$. Let u be the unique weak solution to the Neumann problem $\int_\Omega \langle \nabla u, \nabla h \rangle = \int_{\partial\Omega} f h \, d\sigma$ for all $h \in H_1(D)$ which satisfies $\int_{\partial\Omega} u \, d\sigma = 0$. Then*

$$\|N_\alpha(\nabla u)\|_{L^2(\partial\Omega, d\sigma)} \leqslant C \|f\|_{L^2(\partial\Omega, d\sigma)}$$

where C depends only on D.

Proof. By the method of layer potentials (Section 1) we know that $u \in C^\infty(\overline{\Omega})$. Therefore, we can apply (4.9) to it and obtain

$$\int_{\partial\Omega} |\nabla_t u|^2 \langle Q, N_Q \rangle \, d\sigma(Q)$$
$$= \int_{\partial\Omega} \left\{ f(Q)^2 \langle Q, N_Q \rangle + 2 \langle \alpha(Q), \nabla u \rangle f - (n-2) u f \right\} \, d\sigma(Q).$$

As in the proof of (3.1), $\langle Q, N_Q \rangle \geqslant C > 0$, where c depends only on D. Moreover, $|\langle Q, N_Q \rangle| \leqslant |Q| \leqslant 2 \operatorname{diam} D$, and the tangential vector $\alpha(Q)$ also satisfies $|\alpha(Q)| \leqslant 2 \operatorname{diam} D$. Therefore,

$$\int_{\partial\Omega} |\nabla_t u|^2 \, d\sigma \leqslant C \left\{ \int_{\partial D} f^2 \, d\sigma + \int_{\partial\Omega} |\nabla_t u| \, |f| \, d\sigma + \int_{\partial\Omega} |u| \, |f| \, d\sigma \right\}$$

$$\leqslant C \left\{ \int_{\partial\Omega} f^2 \, d\sigma + \left(\int_{\partial\Omega} |\nabla_t u|^2 \, d\sigma \right)^{1/2} \left(\int_{\partial\Omega} f^2 \, d\sigma \right)^{1/2} \right.$$

$$\left. + \left(\int_{\partial\Omega} u^2 \, d\sigma \right)^{1/2} \left(\int_{\partial\Omega} f^2 \, d\sigma \right)^{1/2} \right\}.$$

Since $\int_{\partial\Omega} u \, d\sigma = 0$, the well-known Poincaré inequality [37] implies that

$$\int_{\partial\Omega} u^2 \, d\sigma \leqslant C \int_{\partial\Omega} |\nabla_t u|^2 \, d\sigma,$$

where C depends on D. Thus if $A = \int_{\partial\Omega} |\nabla_t u|^2 \, d\sigma$ and $B = \int_{\partial\Omega} f^2 \, d\sigma$, we have shown that

$$A \leqslant C(B + A^{1/2}B^{1/2}).$$

We can now employ a standard trick to show that $A \leqslant (2C)^2 B$. First if $A \leqslant B$, then we are done. If $A \geqslant B$, then dividing the inequality above by $A^{1/2}$,

$$A^{1/2} \leqslant C\left(B^{1/2}(B/A)^{1/2} + B^{1/2}\right) \leqslant 2CB^{1/2},$$

and $A \leqslant (2C)^2 B$. Finally, since $|\nabla u|^2 = |\nabla_t u|^2 + f^2$, we have

$$\int_{\partial\Omega} |\nabla u|^2 \, d\sigma \leqslant C \int_{\partial\Omega} f^2 \, d\sigma.$$

Now apply (3.2) to each of the harmonic functions $\partial u / \partial x_j$, $j = 1, \ldots, n$, and the lemma follows.

The rest of the proof of Theorem (4.1) is just a matter of devising an appropriate limiting argument.

(4.11) LEMMA. *Let u_j be a uniformly bounded sequence of elements of $\dot{H}_1(D)$ such that each u_j is harmonic in D. Then there is a subsequence $\{j_k\}$ and a harmonic function $u \in \dot{H}_1(D)$ such that $u_{j_k} \to u$ weakly in $\dot{H}_1(D)$ and $\nabla u_{j_k} \to \nabla u$ uniformly on compact subsets of D. (Once again an element of $\dot{H}_1(D)$ is identified with any of its representatives.)*

Proof. The sequence ∇u_j is uniformly bounded in $L^2(D)$. Let K be a compact subset of D and $d = \operatorname{dist}(K, \partial D)$. For each $X \in K$ the mean value theorem (1.6) implies

$$\nabla u_j(X) = \frac{1}{B(X, d/2)} \int_{B(X, d/2)} \nabla u_j(Y) \, dY.$$

And so

$$|\nabla u_j(X)| \le \left(\frac{1}{B(X, d/2)} \int_{B(X, d/2)} |\nabla u_j(Y)|^2 \, dY \right)^{1/2}$$

and ∇u_j is uniformly bounded on compact subsets of D. Therefore, (see [1]) there is a vector $v = (v_1, \ldots, v_n)$ of harmonic functions and a subsequence j_k such that ∇u_{j_k} converges to v uniformly on compact subsets. Since $\dot{H}_1(D)$ is a Hilbert space, we can also assume that u_{j_k} converges weakly in $\dot{H}_1(D)$ to a function u. It is an easy matter to check, using for instance (1.9) and (4.7), that $\nabla u = v$ and hence that u is harmonic.

Let D be a starlike Lipschitz domain and let Ω_j be smooth domains as above. Let K be a compact subset of D and denote

$$N_\alpha^K(\nabla u)(Q) = \sup\{|\nabla u(X)| : X \in \Gamma_\alpha(Q) \cap K\}.$$

Let $f \in C^\infty(\mathbb{R}^n)$ with $\int_{\partial D} f \, d\sigma = 0$. We will first prove that there is a weak solution u to the Neumann problem in D with boundary data f satisfying

$$\|N_\alpha^K(\nabla u)\|_{L^2(\partial D, d\sigma)} \le C\|f\|_{L^2(\partial D, d\sigma)}$$

with C independent of f and K.

Let $f_j = f - c_j$, where

$$c_j = \frac{1}{\sigma_j(\partial \Omega_j)} \int_{\partial \Omega_j} f \, d\sigma_j.$$

(σ_j denotes surface measure on $\partial \Omega_j$.) Clearly, $c_j \to 0$ as $j \to \infty$, $f_j \in C^\infty(\mathbb{R}^n)$ and $\int_{\partial \Omega_j} f_j \, d\sigma_j = 0$. Also, we can assume

$$\|f_j\|_{L^2(\partial \Omega_j, d\sigma_j)} \le 2\|f\|_{L^2(\partial D, d\sigma)}.$$

Let u_j be the weak solution to the Neumann problem in Ω_j with boundary data f_j such that $\int_{\partial \Omega_j} u_j \, d\sigma_j = 0$. By Remark 3,

$$\|u_j\|_{\dot{H}_1(D)} \le \|u_j\|_{\dot{H}_1(\Omega_j)} \le C\|f_j\|_{L^2(\partial \Omega_j, d\sigma_j)} \le 2C\|f\|_{L^2(d\sigma)}.$$

Thus $\{u_j\}$ is a uniformly bounded sequence in $\dot{H}_1(D)$. By (4.11) we can replace $\{u_j\}$ by a subsequence which we also call $\{u_j\}$ such that $u_j \to u$ weakly in $H_1(D)$. In other words, $\nabla u_j \to \nabla u$ weakly in $L^2(D)$. We claim that u is a weak solution to the Neumann problem in D with boundary data $f|_{\partial D}$. In fact, let $h \in H_1(D)$, then

$$\lim_{j \to \infty} \int_D \langle \nabla u_j, \nabla h \rangle = \int_D \langle \nabla u, \nabla h \rangle.$$

Moreover, $\lim_{j \to \infty} \int_{\Omega_j \setminus D} \langle \nabla u_j, \nabla h \rangle = 0$ because

$$\int_{\Omega_j} |\nabla u_j|^2 \leqslant \|u_j\|_{\dot{H}_1(\Omega_j)} \leqslant C\|f\|_{L^2(d\sigma)}.$$

Therefore,

$$\int_D \langle \nabla u, \nabla h \rangle = \lim_{j \to \infty} \int_{\Omega_j} \langle \nabla u_j, \nabla h \rangle = \lim_{j \to \infty} \int_{\partial \Omega_j} f_j h \, d\sigma_j = \int_{\partial D} fh \, d\sigma,$$

which shows that u is a weak solution.

Next, observe that nontangential regions $\Gamma_\alpha(Q)$ for $Q \in \partial D$ are contained in the corresponding regions for Ω_j. More precisely, there is $\beta > 0$ such that if $Q = (\phi(\theta), \theta)$, $Q_j = (\psi_j(\theta), \theta)$, then for all j,

$$\Gamma_\alpha(Q) = \{X \in D: |X - Q| < (1 + \alpha)\operatorname{dist}(X, \partial D)\} \subset \Gamma_\beta^j(Q_j)$$

$$= \{X \in \Omega_j: |X - Q_j| < (1 + \beta)\operatorname{dist}(X, \partial \Omega_j)\}.$$

Also, by (4.11) $\nabla u_j \to \nabla u$ uniformly on K. Therefore, by Fatou's lemma and (4.10)

$$\int_{\partial D} N_\alpha^K(\nabla u)(Q)^2 \, d\sigma(Q) \leqslant \liminf_{j \to \infty} \int_{\partial \Omega_j} N_\beta(\nabla u_j)(Q)^2 \, d\sigma_j(Q)$$

$$\leqslant \liminf_{j \to \infty} C \int_{\partial \Omega_j} f_j^2 \, d\sigma_j = C \int_{\partial D} f^2 \, d\sigma.$$

Finally, let $f \in L^2(\partial D, d\sigma)$, $\int_{\partial D} f \, d\sigma = 0$. There is a sequence $g_k \in C^\infty(\mathbb{R}^n)$ such that $\int_{\partial D} g_k \, d\sigma = 0$ and $g_k \to f$ in $L^2(d\sigma)$. If u_k is the weak solution to the Neumann problem for data g_k constructed above, then as before there is a harmonic function u such that $u_k \to u$ in $\dot{H}_1(D)$ and u is a weak solution to the Neumann problem in D for data f. Once again we can assume using (4.11) that $\nabla u_k \to \nabla u$ uniformly on compact subsets. Using Fatou's lemma and the corresponding inequality for u_k, we conclude that

$$\|N_\alpha^K(\nabla u)\|_{L^2(\partial D, d\sigma)} \leq C\|f\|_{L^2(\partial D, d\sigma)},$$

with C independent of K. Let K increase to D. We deduce from the monotone convergence theorem the full inequality

$$\|N_\alpha(\nabla u)\|_{L^2(d\sigma)} \leq C\|f\|_{L^2(d\sigma)}.$$

Using the proof of (3.2) we see that $\nabla u(X)$ has nontangential limits $\nabla u(Q)$ a.e. $Q \, d\sigma$ on ∂D. It remains to show that $\langle N_Q, \nabla u(Q) \rangle = f(Q)$ a.e. $d\sigma$. But by dominated convergence, for every $h \in C^\infty(\bar{D})$,

$$\int_{\partial D} \langle N_Q, \nabla u(Q) \rangle h(Q) \, d\sigma(Q) = \lim_{m \to \infty} \int_{\partial D_m} \frac{\partial u}{\partial N_m} h \, d\sigma_m.$$

Furthermore, by Remark 2 the limit also equals $\int_{\partial D} f h \, d\sigma$. Both $\langle N_Q, \nabla u(Q) \rangle$ and f are already known to belong to $L^2(d\sigma)$, and since $C^\infty(\bar{D})$ is dense in $L^2(d\sigma)$ we have $f(Q) = \langle N_Q, \nabla u(Q) \rangle$ a.e. $d\sigma$.

The solution u is unique up to an additive constant because even the weak solution is unique up to an additive constant.

Remark. There are at least two natural ways to specify u uniquely, not just modulo an additive constant. The first is to ask that $\int_{\partial D} u \, d\sigma = 0$. A little extra work shows that the approximation procedure we gave above yields this solution. Another way to specify u is to ask that $u(0) = 0$, in other words, $\int_{\partial D} u(Q) \, d\omega(Q) = 0$.

This solution can be obtained as a weak limit of solutions on smooth domains that vanish at the origin.

Another application of the integral formula (4.9) is to regularity in the Dirichlet problem.

(4.12) *Definition.* We define a Sobolev space on ∂D by

$$H_1(\partial D) = \{ f \in L^2(\partial D, d\sigma): |\nabla_\theta \tilde{f}| \in L^2(d\sigma),$$

where $\tilde{f}(\theta) = f(\phi(\theta), \theta)$ and $\nabla_\theta \tilde{f}$ is taken in the sense of distributions on $S^{n-1}\}$.

Another way to describe $H_1(\partial D)$ is as the closure of functions f that are the restriction to ∂D of $C^\infty(\mathbb{R}^n)$ in the norm

$$\int_{\partial D} |f|^2 \, d\sigma + \int_{\partial D} |\nabla_t f|^2 \, d\sigma.$$

(4.13) THEOREM. *Let $f \in H_1(\partial D)$ and let u be the solution to the Dirichlet problem $\Delta u = 0$ in D, $u = f$ on ∂D, in the sense of (3.2). Then*

$$\|N_\alpha(\nabla u)\|_{L^2(\partial D, d\sigma)} \leqslant C\|f\|_{H_1(\partial D)}.$$

Proof. This time apply (4.9) reversing the roles of $\nabla_t u$ and $\partial u / \partial N$. Arguing as in (4.10) we obtain

$$\|N_\alpha(\nabla u)\|_{L^2(\partial \Omega, d\sigma)} \leqslant C\|\nabla_t u\|_{L^2(\partial \Omega, d\sigma)}.$$

A passage to the limit from smooth domains to Lipschitz domains establishes the theorem.

5. ESTIMATES FOR HARMONIC MEASURE

In this section, the most technical in the article, we give proofs of the basic lemmas concerning the harmonic measure and the kernel function, namely (2.4), (2.6), (2.7), and (2.9). This will complete the proofs of all the results in Sections 2, 3, and 4. The properties of Lipschitz domains that we will use are these:

There exist numbers M and $r_0 > 0$ such that

(1) if $r < r_0$ and $Q \in \partial D$, then there exists a point $A = A_r(Q) \in D$ such that $M^{-1}r < |A - Q| < Mr$ and $M^{-1}r < \text{dist}(A, \partial D)$;

(2) $^C\!D$ satisfies property 1;

(3) if $X_1, X_2 \in D$, $\text{dist}(X_j, \partial D) > \varepsilon$, and $|X_1 - X_2| < 2^k\varepsilon$, then there is a chain of Mk balls $B_1, \ldots, B_{Mk}$ connecting X_1 and X_2 in the sense that X_1 is the center of B_1, X_2 is the center of B_{Mk}, $B_j \cap B_{j+1} \neq \varnothing$, $B_j \subset D$,

$$M^{-1}\text{diam}\, B_j \leqslant \text{dist}(B_j, \partial D) \leqslant M\,\text{diam}\, B_j,$$

and

$$\text{diam}\, B_j \geqslant M^{-1}\min\{\text{dist}(X_1, B_j), \text{dist}(X_2, B_j)\}.$$

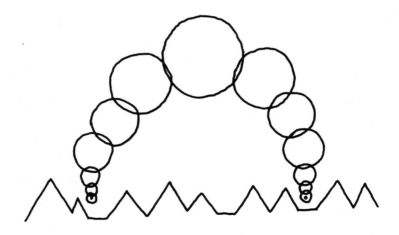

Conditions (1) and (2) follow from the existence for Lipschitz domains of interior and exterior cones. For condition (3) we will indicate the idea of the proof with a picture. The importance of (3) is that it is tailor-made for the application of the argument giving Harnack's inequality. Let P_j denote the center of the ball B_j; B_j and B_{j+1} have comparable diameter and $B_j \cap B_{j+1} \neq \varnothing$. Hence the formula for the Poisson kernel (1.15) in balls slightly larger than B_j and B_{j+1} shows that, if u is a positive harmonic function in D,

$$C_M^{-1} u(P_j) < u(P_{j+1}) < C_M u(P_j),$$

where C_M depends only on M. Therefore, if X_1 and X_2 are as in (3), there is a larger constant R_M, depending only on M, such that $R_M^{-k} u(X_2) < u(X_1) < R_M^k u(X_2)$. In an abuse of notation we will replace R_M by the constant M, since M is just a constant depending on the domain D. Later on we will often replace M by a larger constant, depending only on M, and denote it again by M.

A further property of Lipschitz domains is

(4) Given $r < r_0$ and $Q \in \partial D$, there exists a Lipschitz domain $\Omega_r(Q)$, with Lipschitz constant depending only on D and not Q or r, such that

$$B(Q, r) \cap D \subset \Omega_r(Q) \subset B(Q, Mr) \cap D.$$

To construct $\Omega_r(Q)$ just take an inverted cone like $\mathcal{C}$ (in the proof of (3.16)) and intersect it with D.

Domains with properties (1) (2) and (3) are called *nontangentially accessible* (NTA) domains. (See (6.4) and [28].) P. W. Jones [31] proved that in any NTA domain one can construct $\Omega_r(Q)$ as in (4), where $\Omega_r(Q)$ is NTA instead of Lipschitz. We will now study harmonic measure in Lipschitz domains. Since we will only use properties (1), (2), (3) and (4) the results are also valid in NTA domains.

5.1 LEMMA.   *There exists $\beta > 0$ such that for all $Q \in \partial D$, $r < r_0$ and every positive harmonic function u in D, if u vanishes continuously on $\partial D \cap B(Q, r)$, then for $X \in D \cap B(Q, r)$, $u(X) \leqslant M(|X - Q|/r)^\beta C(u)$ where*

$$C(u) = \sup\{u(Y): Y \in B(Q, r) \cap D\}.$$

Proof. Let $v(X)$ be the harmonic function on $B(Q, r) \cap D$ with boundary values

$$v(X) = \begin{cases} 1 & X \in \bar{D} \cap \partial B(Q, r), \\ 0 & X \in (\partial D) \cap B(Q, r). \end{cases}$$

By the maximum principle, $u(X) \leqslant C(u)v(X)$, so it suffices to show that $v(X) \leqslant M(|X - Q|/r)^{\beta}$. Condition (2) implies that there exists a sequence $\{r_j\}_{j=1}^{\infty}$ such that

$$M^{-4j}r \leqslant r_j \leqslant M^{-4j+2}r,$$

and $\partial B(Q, r_j) \cap {}^C D$ contains at least some fixed fraction of the full surface measure of $\partial B(Q, r_j)$. By the maximum principle, v is dominated by the Poisson integral on the ball $B(Q, r_1)$ of a function that is 1 on $\partial B(Q, r_1) \cap D$ and 0 on $\partial B(Q, r_1) \cap {}^C D$. An easy lower bound on the Poisson kernel for the ball (see (1.15)) yields $v(X) \leqslant 1 - \varepsilon$ for $X \in B(Q, r_2) \cap D$, where $\varepsilon > 0$ depends only on M. Iterating this procedure we obtain $v(X) \leqslant (1 - \varepsilon)^j$ for $X \in B(Q, r_j) \cap D$. In other words, for some $\beta > 0$ depending on M, $v(X) \leqslant M(|X - Q|/r)^{\beta}$ for $X \in B(Q, r) \cap D$.

(5.2) LEMMA. *Let $r < r_0$. If $Q \in \partial D$ and $A_r(Q)$ is given by (1), then*

$$\omega^X(\Delta(Q, 2r)) \geqslant M^{-1} \quad \text{for all} \quad X \in B(A_r(Q), \tfrac{1}{2}M^{-1}r).$$

Proof. Let v be as in (5.1). By the maximum principle $\omega^X(\Delta(Q, 2r)) \geqslant 1 - v(X)$ for $X \in B(Q, r) \cap D$. By property (1) we can choose $A' \in D$ such that $|A' - Q| < M^{-10}r$ and $\mathrm{dist}(A', \partial D) > M^{-20}r$. By (5.1),

$$\omega^{A'}(\Delta(Q, 2r)) \geqslant 1 - v(A') \geqslant \varepsilon.$$

Because A' can be connected to $B(A_r(Q), \tfrac{1}{2}M^{-1}r)$ by a chain of balls as in (3) of length depending only on M, the lemma follows from Harnack's principle.

Let $G(X, Y)$ denote the Green function of D. (Recall that G is negative.) If $r < r_0$, then for $X \in D \setminus B(A_r(Q), \frac{1}{2} M^{-1} r)$

$$r^{n-2} |G(X, A_r(Q))| \leq M \omega^X (\Delta(Q, 2r)). \tag{5.3}$$

This follows from (5.2), the estimate $|G(X, A_r(Q))| \leq C|X - A_r(Q)|^{2-n}$, and the maximum principle in $D \setminus B(A_r(Q), \frac{1}{2} M^{-1} r)$. (The estimate $|G(X, A_r(Q))| \leq C|X - A_r(Q)|^{2-n}$ is valid even when $n = 2$, by comparison with the Green function for the exterior of a ball provided by condition (2).)

(5.4) LEMMA. *If* u *is harmonic in* D, $u \geq 0$, *and* u *vanishes continuously on* $\Delta(Q, 2r)$, *then*

$$u(X) \leq M' u(A_r(Q)) \text{ for all } X \in B(Q, r) \cap D$$

for some M' *depending only on* M.

Proof. Let $Q_0 \in \partial D$. If u vanishes on $\Delta(Q_0, s)$ then by Lemma (5.1), there is M_1 depending only on M so that

$$\sup \{ u(X) : X \in B(Q_0, M_1^{-1} s) \cap D \}$$

$$\leq \frac{1}{2} \sup \{ u(X) : X \in B(Q_0, s) \cap D \}. \tag{5.5}$$

Normalize u so that $u(A_r(Q)) = 1$. By (3) there is a constant M_2 depending only on M_1 such that if $u(Y) > M_2^h$ and $Y \in B(Q, r) \cap D$, then dist$(Y, \partial D) < M_1^{-h} r$. Choose N so that $2^N > M_2$. Finally, let $M' = M_2^h$, where $h = N + 3$.

Suppose that there exists $Y_0 \in B(Q, r) \cap D$ such that $u(Y_0) > M' u(A_r(Q)) = M_2^h$. Then $d(Y_0, \partial D) < M_1^{-h} r$ and if Q_0 is a point of ∂D nearest to Y_0,

$$|Q - Q_0| < r + M_1^{-h} r < 3r/2.$$

Applying (5.5):

$$\sup \{ u(X) : X \in B(Q_0, M_1^{-h+N} r) \cap D \}$$

$$\geq 2^N \sup \{ u(X) : X \in B(Q_0, M_1^{-h} r) \} \geq M_2^{h+1}.$$

Hence, we can choose $Y_1 \in B(Q_0, M_1^{-h+N}) \cap D$ such that $u(Y_1) \geq M_2^{h+1}$. As before for Y_0, $\text{dist}(Y_1, \partial D) < M_1^{-h-1}r$. Let Q_1 be a point of ∂D closest to Y_1. Continuing in this manner we obtain two sequences, $\{Y_k\}$ and $\{Q_k\}$ such that

$$u(Y_k) \geq M_2^{h+k}, \, d(Y_k, \partial D) = |Y_k - Q_k| < M_1^{-h-k}r,$$

and

$$Y_k \in B(Q_{k-1}, M_1^{-h-k+N}) \cap D.$$

This contradicts the continuity of u at $\Delta(Q, 2r)$ provided we can show that the sequence $B(Q_{k-1}, M_1^{-h-k+N}r)$ is contained in $B(Q, 2r)$. In fact,

$$|Y_k - Q| \leq |Y_k - Q_k| + |Q_k - Y_{k-1}| + |Y_{k-1} - Q|$$
$$< \left(M_1^{-h-k} + M_1^{-h-k+N}\right)r + |Y_{k-1} - Q|.$$

Since $|Y_0 - Q| < r$, we have

$$|Y_k - Q| \leq r + \sum_{j=1}^{k} \left(M_1^{-h-j} + M_1^{-h-j+N}\right)r < 2r,$$

because $h = N + 3$.

$$\omega^X(\Delta(Q, r)) \leq M'r^{n-2}|G(X, A_r(Q))| \tag{5.6}$$

for $X \in D \setminus B(Q, 2r)$, $2r < r_0$.

To prove (5.6) fix $X \in D \setminus B(Q, 2r)$ and define

$$g(P) = \begin{cases} G(X, P) & P \in D, \\ 0 & P \in {}^cD. \end{cases}$$

g is continuous in $\mathbb{R}^n \setminus \{X\}$ and subharmonic. For all $P \in \mathbb{R}^n \setminus (\partial D \cup \{X\})$,

$$g(P) = -c_n \left(|X - P|^{2-n} - \int_{\partial D} |Q - P|^{2-n} \, d\omega^X(Q)\right). \tag{5.7}$$

By Fatou's lemma, $\int_{\partial D}|Q - P|^{2-n} d\omega^X(Q) < \infty$ for all $P \in \partial D$. Moreover, if we choose $P_j \to P$ nontangentially, that is, $|P_j - P| \le Md(P_j, \partial D)$, then for all $Q \in \partial D$,

$$|Q - P| \le |Q - P_j| + |P_j - P| \le (M + 1)|Q - P_j|.$$

Therefore, by dominated convergence (5.7) is valid for all $P \in \mathbb{R}^n \backslash \{X\}$.

For any $\phi \in C_0^\infty(\mathbb{R}^n)$ satisfying $\phi(X) = 0$,

$$\int_D g(P)\Delta\phi(P)\, dP$$

$$= -c_n \int_{\mathbb{R}^n} \left\{ |X - P|^{2-n} - \int_{\partial D} |P - Q|^{2-n} d\omega^X(Q) \right\} \Delta\phi(P)\, dP,$$

$$= c_n \int_{\partial D} \int_{\mathbb{R}^n} |P - Q|^{2-n} \Delta\phi(P)\, dP\, d\omega^X(Q),$$

$$= -\int_{\partial D} \phi(Q)\, d\omega^X(Q).$$

Choose $\phi \ge 0$ so that $\phi = 1$ on $\Delta(Q, r)$, ϕ vanishes outside $B(Q, 3r/2)$ and

$$\left| \frac{\partial}{\partial x_i} \frac{\partial}{\partial x_j} \phi \right| \le Mr^{-2}.$$

Then

$$\omega^X(\Delta(Q, r)) \le \int_{\partial D} \phi(Q)\, d\omega^X(Q)$$

$$= -\int_D g(P)\Delta\phi(P)\, dP \le \int_{B(Q,2r)} |g(P)|\, |\Delta\phi(P)|\, dP$$

$$\le M'|G(X, A_r(Q))| r^{n-2}, \text{ by Lemma (5.4)}.$$

Combining (5.6) and (5.3), we obtain

(5.8) LEMMA.   *If $2r < r_0$ and $X \in D \setminus B(Q, 2r)$, then*

$$M^{-1} < \omega^X(\Delta(Q, r))/r^{n-2}|G(X, A_r(Q))| < M.$$

(We have replaced M' by M, since M' is just a constant depending on the previous value of M.)

Observe that in (5.8) the estimate is uniform as X tends to $\partial D \setminus \Delta(Q, 2r)$.

(5.9) LEMMA (*doubling condition*).

$$\omega^X(\Delta(Q, 2r)) \leqslant C_X \omega^X(\Delta(Q, r)).$$

For $2r < r_0$, (5.9) follows from (5.8) and Harnack's principle. For large r it follows from (5.2) and Harnack's principle. This proves (2.5).

(5.10) LEMMA.   *Let r be such that $Mr < r_0$. Suppose that u and v are positive harmonic functions in D vanishing continuously on $\Delta(Q, Mr)$ for some $Q \in \partial D$ and that $u(A_r(Q)) = v(A_r(Q))$. Then*

$$M^{-1} < u(X)/v(X) < M \text{ for all } X \in B(Q, M^{-1}r) \cap D.$$

Proof.   Let Ω be as in (4) such that

$$B(Q, 2M^{-1}r) \cap D \subset \Omega \subset B(Q, \tfrac{1}{2}Mr) \cap D.$$

Denote

$$L_1 = \{P \in \partial\Omega \setminus \partial D : d(P, \partial D) < M^{-1}r\};$$

$$L_2 = \{P \in \partial\Omega : d(P, \partial D) \geqslant M^{-1}r\}.$$

L_2 clearly contains a surface ball (of Ω) of radius comparable to r. Covering L_1 with a finite union of balls of size a small constant times r in order to apply (5.8), we find that $\omega_\Omega^X(L_1 \cup L_2) \leqslant M\omega_\Omega^X(L_2)$ for $X \in B(Q, M^{-1}r) \cap D$. Lemma (5.4) implies $u(X) \leqslant Mu(A_r(Q))$ for $X \in \overline{\Omega}$. By the maximum principle, since u vanishes on $\partial\Omega \cap \partial D$,

$$u(X) \leqslant M\omega_\Omega^X(L_1 \cup L_2)u(A_r(Q)).$$

On the other hand, Harnack's principle and (3) imply $v(X) \geqslant Mv(A_r(Q))$ for all $X \in L_2$. Hence,

$$v(X) \geqslant Mv(A_r(Q))\omega_\Omega^X(L_2)$$

for all $X \in \Omega$, and Lemma (5.10) follows.

(5.11) LEMMA. *Let* $\Delta = \Delta(Q_0, r)$, $r < r_0$. *Let* $\Delta' = \Delta(Q, s) \subset \Delta(Q_0, r/2)$, $Q, Q_0 \varepsilon \partial D$. *If* $X \in D \setminus B(Q_0, 2r)$, *then*

$$\omega^{A_r(Q_0)}(\Delta') \simeq \omega^X(\Delta')/\omega^X(\Delta).$$

($C_1 \simeq C_2$ *means that the ratio of* C_1 *and* C_2 *is bounded above and below by a constant depending only on* M.)

Proof. By (5.8),

$$\omega^X(\Delta) \simeq r^{n-2}|G(X, A_r(Q_0))|,$$

$$\omega^X(\Delta') \simeq s^{n-2}|G(X, A_s(Q))|, \qquad \text{and}$$

$$\omega^{A_r(Q_0)}(\Delta') \simeq s^{n-2}|G(A_r(Q_0), A_s(Q))|.$$

Thus it suffices to prove

$$|G(A_r(Q_0), A_s(Q_0))| \simeq r^{2-n}|G(X, A_s(Q))|/|G(X, A_r(Q_0))|.$$

$$(5.12)$$

Let $u(Y) = G(A_r(Q_0), Y)$; $v(Y) = G(X, Y)$. Choose a point A such that $|A - A_r(Q)| = \frac{1}{2}M^{-1}r$ and $d(A, \partial D) \geqslant \frac{1}{2}M^{-1}r$. Then $u(A) \simeq r^{2-n}$ and $v(A) \simeq |G(X, A_r(Q))|$. Now apply (5.10) to appropriate multiples of u and v, and let $Y = A_s(Q)$ to obtain (5.12). This implies (2.7).

Recall the kernel function $K(A, Q)$ defined in (2.8). The doubling condition (5.9) implies

$$K(A, Q) = \lim \frac{\omega^A(\Delta')}{\omega(\Delta')} \text{ a.e. } (\omega),$$

as Δ' shrinks to Q. A priori, $K(A, Q)$ is only defined for almost every $(\omega)Q$. Actually, $K(A, Q)$ is a Hölder continuous function of Q as we will see later.

(5.13) LEMMA. Let $A = A_r(Q_0)$, $Q_0 \in \partial D$, $\Delta_j = \Delta(Q_0, 2^j r)$ and $R_j = \Delta_j \setminus \Delta_{j-1}$. Then

$$\sup\{K(A, Q): Q \in R_j\} \leqslant c_j / \omega(\Delta_j),$$

with $c_j \leqslant CM2^{-\alpha j}$; $\alpha > 0$, M depend only on D. (C depends only on the choice of X_0.)

Proof. First consider j such that $2^j r < r_0$. Pick a small surface ball $\Delta' \subset R_j$. Denote $A_j = A_{2^j r}(Q_0)$. By (5.11),

$$\omega^{A_j}(\Delta') \simeq \omega(\Delta') / \omega(\Delta_j).$$

By Lemmas (5.1) and (5.4),

$$\omega^A(\Delta') \leqslant M\omega^{A_j}(\Delta') \left(\frac{|A - Q_0|}{2^j r} \right)^\beta \leqslant M \frac{\omega(\Delta')}{\omega(\Delta_j)} 2^{-j\beta}.$$

Hence,

$$\frac{\omega^A(\Delta')}{\omega(\Delta')} \leqslant \frac{M2^{-j\beta}}{\omega(\Delta_j)}, \text{ and } c_j \leqslant M2^{-j\beta}.$$

There is only a finite number of j for which R_j is nonempty and $2^j r > r_0$. Thus it is enough to show that

$$\sup\{K(A, Q): Q\varepsilon\partial D \setminus \Delta(Q_0, r_0)\} \leqslant C.$$

Choose $\Delta' \subset \partial D \backslash \Delta(Q_0, r_0)$. By (5.4) and Harnack's inequality

$$\omega^A(\Delta') \leq M \omega^{A_{r_0}(Q_0)}(\Delta') \leq M^2 \omega(\Delta').$$

Note that (2.6) is an immediate consequence of this Lemma and (3).

(5.14) LEMMA.   *Let* $\Delta = \Delta(Q_0, r)$, $r < r_0$. *Then*

$$\sup\{K(X, Q): Q \in \partial D \backslash \Delta\} \to 0 \ as \ X \to Q_0.$$

Proof.   Let Δ' be a small surface ball about Q. As in the proof of (5.13), $\omega^x(\Delta') \leq M^2 \omega(\Delta')$ for $X \in B(Q_0, r/2) \cap D$. Since $\omega^x(\Delta')$ vanishes continuously on $B(Q_0, r/2) \cap \partial D$ we deduce from (5.1) the stronger estimate:

$$\omega^X(\Delta') \leq M^3 \omega(\Delta')\big(|X - Q_0| r^{-1}\big)^\beta.$$

Thus

$$K(X, Q) \leq M^3 \big(|X - Q_0| r^{-1}\big)^\beta,$$

and the lemma follows.

All that remains to be established is (2.9). Our starting point is a different notion of kernel function that will turn out to be the same as the one in (2.8).

Definition.   A *kernel function* u in D at $Q \in \partial D$ is a positive harmonic function in D that vanishes continuously on $\partial D \backslash \{Q\}$ and such that $u(X_0) = 1$. (X_0 is fixed as in (2.8); $\omega = \omega^{X_0}$.)

(5.15) LEMMA.   *Let D be a Lipschitz domain. Let u be positive in D, continuous in $\overline{D} \backslash \{Q\}$ where $Q \in \partial D$. If $u = 0$ on $\partial D \backslash \Delta(Q, r)$, then for all $X \in D \backslash B(Q, Mr)$,*

$$u(X) \simeq u(A_r(Q)) \omega^X(\Delta(Q, r)).$$

Proof. Cover $\partial B(Q, Mr) \cap \partial D$ with a finite collection of surface balls of ∂D of size roughly r, disjoint from $\Delta(Q, r)$. Both $u(X)$ and $u(A)\omega^X(\Delta(Q, r))$ (with $A = A_r(Q)$) vanish continuously on these surface balls. By (5.10), (5.2), and Harnack's inequality, we obtain the desired estimate for all $X \in \bar{D} \cap \partial B(Q, Mr)$. The full estimate then follows from the maximum principle.

(5.16) LEMMA. *Let D be a Lipschitz domain and $Q \in \partial D$. Then there exists a kernel function in D at Q.*

Proof. Let r_j be a positive sequence tending to zero. Denote

$$u_j(X) = \omega^X\big(\Delta(Q, r_j)\big) / \omega\big(\Delta(Q, r_j)\big).$$

Obviously $u_j(X_0) = 1$. By Harnack's inequality the sequence u_j is uniformly bounded on compact subsets of D. Therefore, there exists a subsequence, denoted again by u_j such that u_j tends to a positive harmonic function u uniformly on compact subsets of D. By (5.15), if $r > r_j$, then

$$u_j(X) \leqslant M u_j(A_r(Q)) \omega^X(\Delta(Q, r)).$$

Letting j tend to infinity, we conclude that $u(X)$ vanishes continuously on $\partial D \backslash \Delta(Q, 2r)$ for any $r > 0$. Clearly, $u(X_0) = 1$ and therefore u is a kernel function for D at Q.

(5.17) LEMMA. *Suppose that u_1 and u_2 are two kernel functions for D at Q. Then*

$$M^{-1} < u_1(X)/u_2(X) < M \text{ for all } X \in D.$$

Proof. Since $u_1(X_0) = 1$, (5.15) implies $u_1(A_r(Q))\omega(\Delta(Q, r)) \simeq 1$ for all $r > 0$. Using (5.15) again at X, we have

$$u_1(X) \simeq \omega^X(\Delta(Q, r))/\omega(\Delta(Q, r))$$

for all $X \in D \setminus B(Q, Mr)$. Choose a sequence r_j as in (5.16) so that

$$u(X) = \lim_{j \to \infty} \frac{\omega^X(\Delta(Q, r_j))}{\omega(\Delta(Q, r_j))}$$

uniformly on compact subsets. Then $u_1(X) \simeq u(X)$. Similarly $u_2(X) \simeq u(X)$ and the lemma follows.

(5.18) THEOREM. *Let D be a Lipschitz domain. For every $Q \in \partial D$ there exists a unique kernel function for D at Q.*

Proof. We need only prove uniqueness. Let u_1 and u_2 be two kernel functions at Q. By (5.16),

$$M^{-1} \leqslant u_1(X)/u_2(X) \leqslant M \text{ for all } X \in D.$$

If $M = 1$, then we are done. Suppose that $M > 1$. Let $A = 1/(M - 1)$. Then $u_1 + A(u_1 - u_2)$ is also a kernel function at Q. From (5.16) we find $u_2 \leqslant M(u_1 + A(u_1 - u_2))$. Therefore,

$$\left[u_1 + A(u_1 - u_2) \right] + A\left[(u_1 - u_2) + A(u_1 - u_2) \right]$$

is a kernel function at Q. Proceeding inductively, we conclude that there is a sequence of numbers A_j tending to infinity such that $u_1 + A_j(u_1 - u_2)$ is a kernel function at Q. Thus by (5.16), $u_1 + A_j(u_1 - u_2) \leqslant Mu_1$. This can only be true for all j if $u_1 \leqslant u_2$. Similarly, $u_1 \geqslant u_2$.

(5.19) COROLLARY. *For each $Q \in \partial D$, $\lim_{r \to 0} \dfrac{\omega^X(\Delta(Q, r))}{\omega(\Delta(Q, r))}$ exists.*

Proof. It is enough to show that for any sequence $r_m \to 0$ there is a subsequence r_{m_j} such that

$$\lim_{j \to \infty} \frac{\omega^X(\Delta(Q, r_{m_j}))}{\omega(\Delta(Q, r_{m_j}))} = u(X),$$

where u is the kernel function for D at Q. This follows from the proof of (5.16) and (5.18).

Denote

$$K(X, Q) = \lim_{r \to 0} \frac{\omega^X(\Delta(Q, r))}{\omega(\Delta(Q, r))}.$$

We have shown that this exists for every Q. Clearly,

$$K(X, Q) = \frac{d\omega^X}{d\omega}(Q),$$

the Radon-Nykodym derivative. At this point the continuity of $K(X, Q)$ as a function of Q follows easily from (5.18). We prefer to give a direct estimate on its modulus of continuity which shows that it is Hölder continuous.

(5.20) THEOREM. *There is a constant M depending only on the Lipschitz constant of D such that if $Q_0, Q_1 \in \partial D$, $X \in D$ and $|X - Q_0| \geq M^j |Q_1 - Q_0|$, then*

$$\left| \frac{K(X, Q_1)}{K(X, Q_0)} - 1 \right| \leq M(1 - M^{-1})^j.$$

COROLLARY: *$K(X, Q)$ is Hölder continuous as a function of Q: $|K(X, Q) - K(X, Q')| < C_X |Q - Q'|^\alpha$ for some $\alpha > 0$, depending only on the Lipschitz constant of D.*

The corollary is an immediate consequence of (5.20) if we multiply the inequality in (5.20) by $K(X, Q_0)$.

Proof of (5.20):

(5.21) LEMMA. *Let $M > 2$. Let μ be a positive finite measure on a set S, and let θ be a measurable function on S such that $0 < a \leq \theta \leq A$.*

Denote

$$B(\theta) = \sup\left\{\frac{\displaystyle\int_S \theta(x)w(x)\,d\mu(x)}{\displaystyle\int_S w(x)\,d\mu(x)} : M^{-1} < w < M\right\}$$

$$b(\theta) = \inf\left\{\frac{\displaystyle\int_S \theta(x)w(x)\,d\mu(x)}{\displaystyle\int_S w(x)\,d\mu(x)} : M^{-1} < w < M\right\}.$$

Then

$$\frac{B(\theta)}{b(\theta)} - 1 \leqslant \left(1 - \frac{1}{4}M^{-2}\right)\left(\frac{A}{a} - 1\right).$$

Proof. Without loss of generality, we may assume that $a = 1$ and $\mu(S) = 1$. Denote $\gamma = (1 - \frac{1}{4}M^{-2})(A - 1)$.

Case 1. $B(\theta) \leqslant 1 + \gamma$.
Observe that $b(\theta) \geqslant 1 = a$. Then

$$\frac{B(\theta)}{b(\theta)} - 1 \leqslant \frac{1+\gamma}{b(\theta)} - 1 \leqslant (1+\gamma) - 1 = \gamma.$$

Case 2. $B(\theta) > 1 + \gamma$.
Denote $S_1 = \{x \in S: \theta(x) > 1 + \frac{1}{2}\gamma\}$. Let $m = \mu(S_1)$. Define

$$\theta_1(x) = \begin{cases} A & x \in S_1 \\ 1 + \frac{1}{2}\gamma & x \in S \setminus S_1. \end{cases}$$

Since $\theta \leqslant \theta_1$, $B(\theta) \leqslant B(\theta_1)$. It is easy to calculate that

$$B(\theta_1) = \frac{AMm + (1 + \frac{1}{2}\gamma)(1 - m)M^{-1}}{Mm + (1 - m)M^{-1}}.$$

Thus

$$1 + \gamma < \frac{AMm + (1 + \frac{1}{2}\gamma)(1 - m)M^{-1}}{Mm + (1 - m)M^{-1}}.$$

Therefore, $\frac{1}{2}\gamma < ((A - 1)M^2 + \frac{1}{2}\gamma - \gamma M^2)m$, and hence $m > 1/2$. Denote

$$\theta_2(x) = \begin{cases} 1 + \frac{1}{2}\gamma & x \in S_1, \\ 1 & x \in S \setminus S_1. \end{cases}$$

$\theta_2 \leqslant \theta$ implies $b(\theta_2) \leqslant b(\theta)$. It is easy to calculate

$$b(\theta_2) = \frac{M^{-1}(1 + \frac{1}{2}\gamma)m + M(1 - m)}{M^{-1}m + M(1 - m)} > \frac{(1 + \frac{1}{2}\gamma)M^{-1} + M}{M^{-1} + M}$$

since $m > 1/2$. Observe that $B(\theta) \leqslant A$. Therefore

$$\frac{B(\theta)}{b(\theta)} - 1 \leqslant \frac{A(M^{-1} + M)}{(1 + \frac{1}{2}\gamma)M^{-1} + M} - 1$$

$$= (A - 1)\left(1 - \frac{1 - \frac{1}{4}M^{-2}}{2M(M + M^{-1})}\right) < \gamma.$$

(5.22) LEMMA. *For any $Q \in \partial D$, $r < r_0$, there exists a Lipschitz domain $\Omega \subset D$ such that $D \setminus B(Q, r) \supset \Omega \supset D \setminus B(Q, Mr)$, and the Lipschitz constant of Ω depends only on the one of D.*

Proof. Remove from D an inverted cone as in $\mathcal{C}$ in the proof of (3.16). (This lemma remains valid in NTA domains if we ask only that Ω be NTA.)

Let $Q_0, Q_1 \in \partial D$; $r = |Q_1 - Q_0|$. Denote by Ω_j the region of (5.22) corresponding to $Q_0 \ldots$ and the distance $M^j r$, $j = 1, 2, \ldots$; $M^j r < r_0$. Replacing M by $2M$ we see that

$$D \setminus B(Q_0, M^j r) \subset \Omega_j \subset D \setminus B(Q_0, 2M^{j-1}r).$$

Let $S_j = \partial \Omega_j \setminus \partial D$. We can assume that $X_0 \in \Omega_j$ for all j ($M^j r < r_0$) and dist($X_0, \partial \Omega_j$) ≈ dist($X_0, \partial D$). Let ω_j^X be harmonic measure for Ω_j and let

$$K_j(X, Q) = \frac{d\omega_j^X}{d\omega_j^{X_0}}(Q)$$

be the kernel function of Ω_j at Q. By (5.11) and the proof of (5.19),

$$M^{-1} < K(A, Q')/\omega(\Delta) < M$$

where $A = A_r(Q_0)$, $\Delta = \Delta(Q_0, r)$ and $Q' \in \Delta$. Hence, replacing M with a larger value, e.g., M^2,

$$M^{-1} < K(A, Q_1)/K(A, Q_0) < M.$$

By (5.10), again replacing M by a larger value,

$$M^{-1} < K(X, Q_1)/K(X, Q_0) < M \text{ for all } X \in D \setminus B(Q_0, 2r).$$

$$(5.23)$$

Similarly,

$$M^{-1} < K_j(X, Q')/K_j(X, Q'') < M \qquad (5.24)$$

for all $X \in D \setminus B(Q_0, 2M^j r)$, $Q', Q'' \in S_j$. Let $u_0(X) = K(X, Q_0)$; $u_1(X) = K(X, Q_1)$. Define

$$b_j = \inf\left\{ \frac{u_1(X)}{u_0(X)} : X \in S_j \right\}, \qquad B_j = \sup\left\{ \frac{u_1(X)}{u_0(X)} : X \in S_j \right\}.$$

The maximum principle implies that $b_j u_0(X) \leqslant u_1(X) \leqslant B_j u_0(X)$ for all $X \in \Omega_j$. Since $u_0(X_0) = u_1(X_0) = 1$, we see that $b_j \leqslant 1 \leqslant B_j$. Next, (5.23) implies that $b_1 > M^{-1}$, $B_1 < M$.

Define $\theta(Q) = u_1(Q)/u_0(Q)$ for $Q \in S_j$, and $d\mu(Q) = u_0(Q) d\omega_j(Q)$ for $Q \in S_j$. Fix $Q_2 \in S_j$ and $X \in D \setminus B(Q_0, 2M^j r)$.

Let

$$w(Q) = K_j(X, Q)/K_j(X, Q_2)$$

for $Q \in S_j$. By (5.24), $M^{-1} < w(Q) < M$ independent of X. Now, since u_0 and u_1 are continuous in $\overline{\Omega}_j$, we can write them as

$$u_1(X) = K_j(X, Q_2) \int_{S_j} \theta(Q) w(Q) \, d\mu(Q),$$

$$u_0(X) = K_j(X, Q_2) \int_{S_j} w(Q) \, d\mu(Q).$$

Therefore,

$$B_{j+1} = \sup \left\{ \frac{u_1(X)}{u_0(X)} : X \in S_{j+1} \right\}$$

$$\leqslant \sup \left\{ \frac{\int_{S_j} \theta(Q) w(Q) \, d\mu(Q)}{\int_{S_j} w(Q) \, d\mu(Q)} : M^{-1} < w < M \right\} = B(\theta),$$

in the notation of (5.21). Similarly, $b_{j+1} \geqslant b(\theta)$. Consequently, since $b_j \leqslant \theta(Q) \leqslant B_j$, (5.21) implies

$$\left(\frac{B_{j+1}}{b_{j+1}} - 1 \right) \leqslant \left(1 - \frac{1}{4} M^{-2} \right) \left(\frac{B_j}{b_j} - 1 \right).$$

By induction (recalling that $b_1 > M^{-1}$ and $B_1 < M$)

$$\left| \frac{B_j}{b_j} - 1 \right| \leqslant M^2 (1 - \varepsilon)^j, \quad \text{with } \varepsilon = \frac{1}{4} M^{-2}.$$

Finally, $b_j \leqslant 1 \leqslant B_j$ and $b_j \leqslant u_1(X)/u_0(X) \leqslant B_j$ for $X \in S_j$ imply that

$$\left| \frac{u_1(X)}{u_0(X)} - 1 \right| \leq M^2(1 - \varepsilon)^j \quad \text{for } X \in S_j.$$

By the maximum principle, the same inequality holds for $X \in \Omega_j$, and (5.20) is proved.

An easy consequence of (5.10) and Harnack's inequality is the following boundary Harnack principle.

(5.25) THEOREM. *Let D be a Lipschitz domain and let V be an open set in $\mathbb{R}^n$. For any compact set $K \subset V$, there exists a constant C such that for all positive harmonic functions u and v in D that vanish continuously on $(\partial D) \cap V$ and satisfy $u(A) = v(A)$ for some $A \in K \cap D$,*

$$C^{-1}u(X) < v(X) < Cu(X) \text{ for all } X \in K \cap \overline{D}.$$

This theorem can be improved using the methods of (5.20) as follows:

(5.26) THEOREM. *Let D, K, V, u, v be as in (5.25). Then there exists $\alpha > 0$ independent of u and v such that $u(X)/v(X)$ is Hölder continuous of order α in $K \cap \overline{D}$. In particular, $\lim_{X \to Q} u(X)/v(X)$ exists for every $Q \in \partial D \cap K$.*

6. FURTHER RESULTS

1. The Fatou type theorem (2.3) has a local analogue. Let $F \subset \partial D$. We say that the harmonic function u is nontangentially bounded from below on F if given $Q \in \partial D$ there exist $\alpha > 0$ and M, which may depend on Q such that $u(X) > -M$ for all $X \in \Gamma_\alpha(Q)$. The local theorem is that if u is nontangentially bounded from below on F, then u has finite nontangential limits at almost every $(d\omega)$ point of F. This result can be reduced to (2.3) easily by considering so-called sawtooth domains: $U_{Q \in F'} \Lambda(Q)$, where $\Lambda(Q)$

is an interior cone at Q. (See [24].) For the corresponding result for smooth domains in $\mathbb{R}^n$, see A. P. Calderón [5] and L. Carleson [9].

2. Another way to characterize the existence of limits of harmonic functions is in terms of the so-called area integral:

$$A_\alpha(u)(Q)^2 = \int_{\Gamma_\alpha(Q)} \text{dist}(X, \partial D)^{2-n} |\nabla u(X)|^2 \, dX.$$

THEOREM (Dahlberg [14]). *Let $F \subset \partial D$ and let u be harmonic in D, a Lipschitz domain. Then u has finite nontangential limits a.e. $(d\omega)$ on F if and only if for almost every Q $(d\omega)$ of F there exists α such that $A_\alpha(u)(Q) < \infty$.*

This theorem was proved in the disk in $\mathbb{R}^2$ by Marcinkiewicz and Zygmund [36] and Spencer [42] and for smooth domains in higher dimensions by A. P. Calderón [6] and E. M. Stein [43].

A global version of the area integral theorem, proved also by Dahlberg [14], deserves further attention. It says

$$\|A_\alpha(u)\|_{L^p(\partial D, d\sigma)} \simeq \|N_\alpha(u)\|_{L^p(\partial D, d\sigma)}$$

in a Lipschitz domain D, for $0 < p < \infty$. (Here u is a harmonic function with $u(X_0) = 0$ for some fixed $X_0 \in D$.) In particular, if $p = 2$, this estimate and Corollary (3.2) imply that if $f \in L^2(\partial D, d\sigma)$ and $\int_{\partial D} f \, d\omega = 0$, then

$$\int_D \text{dist}(X, \partial D) |\nabla u(X)|^2 \, dX \simeq \int_{\partial D} f(Q)^2 \, d\sigma(Q). \qquad (*)$$

For global estimates in smooth domains see Burkholder and Gundy [2], Fefferman and Stein [20], Gundy and Wheeden [23], and Stein [44].

3. The Sobolev space $H_s(D)$ was defined in (4.2). For $s > 0$, define $H_s(\partial D) = $ Restriction of $H_{s+1/2}$ to ∂D. Let $H_0(\partial D) = L^2(\partial D)$. The space $H_1(\partial D)$ coincides with the one given in (4.12),

even on Lipschitz domains. If D is a C^∞ domain, then for all $s \geq 0$, $f \in H_s(\partial D)$ implies $u \in H_{s+1/2}(D)$ where u solves the Dirichlet problem $\Delta u = 0$ and $u|_{\partial D} = f$ (in an appropriate sense). If $g \in H_s(\partial D)$, $s \geq 0$, and v solves the Neumann problem for g ($\Delta v = 0$ and $\partial v / \partial N = g$ on ∂D in the appropriate sense) then $v \in H_{s+3/2}(D)$. A proof of these estimates can be found in [21]. They are the best possible in that it is false in general that $u \in H_{s+1/2+\varepsilon}(D)$ or that $v \in H_{s+3/2+\varepsilon}(D)$ for any $\varepsilon > 0$. On a Lipschitz domain these best possible estimates are valid for $s = 0$. In the Dirichlet problem, if $f \in H_0(\partial D) = L^2(\partial D, d\sigma)$, then u satisfies ($*$) and a characterization of $H_{1/2}(D)$ in terms of a notion called real interpolation [35] shows that $u \in H_{1/2}(D)$. Similarly, if v solves the Neumann problem for $g \in L^2(\partial D, d\sigma)$, then, as we saw in Theorem (4.1), $\partial v / \partial x_j$ has boundary data in $L^2(\partial D, d\sigma)$, so that $\partial v / \partial x_j \in H_{1/2}(D)$, $j = 1, \ldots, n$, or in other words $v \in H_{3/2}(D)$.

4. As we mentioned in Section 5, domains satisfying (1), (2), and (3) of that section are called nontangentially accessible (NTA) domains. The results of Section 5, Theorem (2.3) and its local analogues in 1 and 2 above are valid in NTA domains with almost the same proofs [28]. However, the boundary of an NTA domain can be quite wild. It need not be rectifiable nor have a tangent plane at any point. In fact, given $\beta < n$, there is an NTA domain in $\mathbb{R}^n$ whose boundary has positive β-Hausdorff measure. Examples of NTA domains are Zygmund domains (i.e., domains given locally by graphs of functions $\phi \colon \mathbb{R}^{n-1} \to \mathbb{R}$ satisfying the Zygmund condition $|\phi(X+t) + \phi(x-t) - 2\phi(x)| \leq C|t|$) and quasispheres (i.e., images under a global quasiconformal mapping of $\mathbb{R}^n$ to itself of the unit ball).

5. Many of the results of this article extend to solutions of elliptic operators in divergence form

$$\mathcal{L} = \sum_{i,j} \frac{\partial}{\partial x_i} a_{ij}(x) \frac{\partial}{\partial x_j}$$

satisfying

$$c_1 |\xi|^2 \leq \sum_{i,j} a_{ij}(x) \xi_i \xi_j \leq c_2 |\xi|^2; \qquad a_{ij} = a_{ji}.$$

If the coefficients $a_{ij}(x)$ belong to $C^\infty(\mathbb{R}^n)$, then all of the results above are valid. If the coefficients are merely bounded and measurable, then Theorem (2.3) and the results of Section 5 are valid even on NTA domains ([4], [28]). However the results of Section 3 fail in general for these operators. One can find operators $\mathcal{L}$ (even with continuous coefficients) such that surface measure and the analogue of harmonic measure are mutually singular in the ball in $\mathbb{R}^n$ ([3]).

6. The qualitative results of Section 3 (that is, all but the inequalities) extend to domains with much rougher boundaries than Lipschitz domains. An L_1^p domain in $\mathbb{R}^n$ is a domain whose boundary is given locally in some C^∞ coordinate system as the graph of a continuous function $\phi: \mathbb{R}^{n-1} \to \mathbb{R}$, with $\nabla\phi \in L^p(\mathbb{R}^{n-1})$. Note that L_1^∞ domains are Lipschitz domains. If D is an L_1^p domain with $p > n - 1$, then harmonic measure and surface measure are mutually absolutely continuous. Also, one can solve the Dirichlet problem with boundary data in $L^r(\partial D, d\sigma)$ for r sufficiently large, depending on p. The solution converges nontangentially a.e. $d\sigma$ to its boundary values. There is also an endpoint result when $p = n - 1$. (See [27].)

7. Several notions of Hardy spaces have been considered in C^1, Lipschitz, and NTA domains. (See [33], [34], [17], [15], [18], [28], [30].) For example, on an NTA domain D, define

$$H^p(D, d\omega) = \{u: \Delta u = 0 \text{ in } D, N_\alpha(u) \in L^p(\partial D, d\omega)\}.$$

Using the Hölder continuity of the kernel function (5.20), Jerison and Kenig [28] gave a satisfactory description known as an atomic decomposition of the generalized functions that are boundary values of function in $H^p(D, d\omega)$, for $p_0 < p \le \infty$. (p_0 depends only on D and $0 < p_0 < 1$. The range $p > 1$ is already taken care of in Theorem (2.3$'$). Indeed, $H^p(D, d\omega) = L^p(\partial D, d\omega)$ for $p > 1$. The interest here is the range $p_0 < p \le 1$.) In particular, one can prove that the dual of $H^1(D, d\omega)$ is $\mathrm{BMO}(\partial D, d\omega)$.

On a Lipschitz domain D, one can also consider

$$H^1(D, d\sigma) = \{u: \Delta u = 0 \text{ in } D, N_\alpha(u) \in L^1(\partial D, d\sigma)\}.$$

B. E. J. Dahlberg [15] has derived the atomic decomposition for this space and proved that its dual is a certain weighted BMO class.

A third notion of a Hardy space, based on the Stein-Weiss definition [45] is more closely related to the Neumann problem. Let

$$H^1_{\mathfrak{N}}(D, d\sigma) = \{\vec{u}: \vec{u} = \nabla U, \text{ where } \Delta U = 0$$

$$\text{in } D \text{ and } N_\alpha(|\vec{u}|) \in L^1(\partial D, d\sigma)\}.$$

On C^1 domains D, using the method of layer potentials, Fabes and Kenig [17] gave the atomic decomposition for the space of

$$f = \langle N_Q, \vec{u}(Q) \rangle = \left. \frac{\partial U}{\partial N} \right|_{\partial D},$$

where $\vec{u} \in H^1_{\mathfrak{N}}(D, d\sigma)$. Finally in [18], Fabes, Kenig, and Neri provided a connection between this "Neumann" space and $H^1(D, d\omega)$ for C^1 domains D: $f = \langle N_Q, \vec{u}(Q) \rangle$ for some $\vec{u} \in H^1_{\mathfrak{N}}(D, d\sigma)$ if and only if kf is the boundary value of a function $v \in H^1(D, d\omega)$, where $k = d\omega/d\sigma$. The problem of understanding $H^1_{\mathfrak{N}}(D, d\sigma)$ on Lipschitz domains is open.

8. A more detailed analysis of the questions treated here, and other related results, can be performed in two dimensions by means of conformal mapping. (See [33], [30], [7], [11].)

NOTES

Section 1. We found the works of Kellogg [32], Brelot [1], and Folland [21] particularly useful in preparing this exposition. The reader can consult these works for further references.

Section 2. Theorem (2.1) is due to Brelot and Theorem (2.3) to Hunt and Wheeden [24] and Carleson [9].

Section 3. The proof of Dahlberg's Theorem (3.1) [12] presented here is due to Jerison and Kenig [26]. Formula (3.3) appeared independently much earlier in the work of Payne and Weinberger [40]. The same proof of (3.1) (a) was also discovered independently by K. O. Widman (unpublished; personal communication from the editor).

Section 4.   Theorem (4.1) is due to Jerison and Kenig [29] and is based on the work of Payne and Weinberger [40]. The crucial formula (4.9) was singled out by Payne and Weinberger among a class of similar formulas due to F. Rellich [41].

Section 5.   (5.1)–(5.19) are due to Carleson [9] and Hunt and Wheeden [24] (except for Lemma (5.8), which is due to Dahlberg [12]). However the general strategy of the proof is that of Caffarelli, et al. [4]. The approach is more flexible in that Caffarelli, et al., applied it to divergence class operators with bounded measurable coefficients and Jerison and Kenig [28] applied it to nontangentially accessible domains. The Hölder continuity (5.20) (as opposed to mere continuity) of the kernel function is due to Jerison and Kenig [28].

REFERENCES

1. M. Brelot, "Éléments de la Théorie Classique du Potentiel," *Centre de Documentation Universitaire*, Paris V (1965).
2. D. Burkholder and R. Gundy, "Distribution function inequalities for the area integral," *Studia Math.*, **44** (1972), 527–544.
3. L. Caffarelli, E. Fabes, and C. Kenig, "Completely singular elliptic-harmonic measures," to appear, *Indiana Univ. Math. J.*
4. L. Caffarelli, E. Fabes, S. Mortola, and S. Salsa, "Boundary behavior of non-negative solutions of elliptic operators in divergence form," to appear, *Indiana Univ. Math. J.*
5. A. P. Calderón, "On the behavior of harmonic functions near the boundary," *Trans. Amer. Math. Soc.*, **68** (1950), 47–54.
6. ———, "On a theorem of Marcinkiewicz and Zygmund," *Trans. Amer. Math. Soc.*, **68** (1950), 55–61.
7. ———, "On the Cauchy integral on Lipschitz curves, and related operators," *Proc. Nat. Acad. Sci.*, **74** (1977), 1324–1327.
8. A. P. Calderón, C. P. Calderón, E. Fabes, M. Jodeit Jr., and N. Rivière, "Applications of the Cauchy integral on Lipschitz curves," *Bull. Amer. Math. Soc.*, **84** (1978), 287–290.
9. L. Carleson, "On the existence of boundary values for harmonic functions in several variables," *Ark. Mat.*, **4** (1962), 393–399.
10. R. Coifman and C. Fefferman, "Weighted norm inequalities for maximal functions and singular integrals," *Studia Math.*, **51** (1974), 241–250.
11. R. Coifman and Y. Meyer, "Le théorème de Calderón par les méthodes de variable réelle," *C. R. Acad. Sci. Paris, Ser. A*, t. **289** (1979), 425–428.
12. B. Dahlberg, "Estimates of harmonic measure," *Arch. Rational Mech. Anal.*, **65** (1977), 275–288.
13. ———, "On the Poisson integral for Lipschitz and C^1 domains," *Studia Math.*, **66** (1979), 13–24.
14. ———, "Weighted norm inequalities for the Lusin area integral and the non-tangential maximal functions for functions harmonic in a Lipschitz domain," to appear, *Studia Math.*

15. ———, "A note on H^1 and BMO," preprint.
16. E. Fabes, M. Jodeit Jr., and N. Rivière, "Potential techniques for boundary value problems on C^1 domains," *Acta Math.*, **141** (1978), 165–186.
17. E. Fabes and C. Kenig, "On the Hardy space H^1 of a C^1 domain," to appear, *Ark. Mat.*
18. E. Fabes, C. Kenig, and U. Neri, "Carleson measures, H^1 duality and weighted BMO in non-smooth domains," to appear, *Indiana Univ. Math. J.*
19. P. Fatou, "Séries trigonométriques et séries de Taylor," *Acta Math.*, **30** (1906), 335–400.
20. C. Fefferman and E. Stein, "H^p spaces of several variables," *Acta Math.*, **129** (1972), 137–193.
21. G. Folland, *Introduction to Partial Differential Equations*, Princeton University Press, Princeton, NJ, 1976.
22. F. Gehring, "The L^p integrability of the partial derivatives of a quasiconformal mapping," *Acta Math.*, **139** (1973), 265–277.
23. R. Gundy and R. Wheeden, "Weighted integral inequalities for the non-tangential maximal functions, Lusin area integral and Walsh-Paley series," *Studia Math.*, **49** (1974), 107–124.
24. R. Hunt and R. Wheeden, "On the boundary values of harmonic functions," *Trans. Amer. Math. Soc.*, **132** (1968), 307–322.
25. R. Hunt and R. Wheeden, "Positive harmonic functions on Lipschitz domains," *Trans. Amer. Math. Soc.*, **147** (1970), 507–527.
26. D. Jerison and C. Kenig, "An identity with applications to harmonic measure," *Bull. Amer. Math. Soc.*, vol. 2 (1980), 447–451.
27. D. Jerison and C. Kenig, "The Dirichlet problem in nonsmooth domains," to appear, *Ann. of Math.*
28. D. Jerison and C. Kenig, "Boundary behavior of harmonic functions in nontangentially accessible domains," to appear, *Adv. in Math.*
29. D. Jerison and C. Kenig, "The Neumann problem on Lipschitz domains," to appear, *Bull. Amer. Math. Soc.*
30. D. Jerison and C. Kenig, "Hardy spaces, A_∞ and singular integrals on chord-arc domains," to appear, *Math. Scand.*
31. P. Jones, "Quasiconformal mappings and extendability of functions in Sobolev spaces," to appear, *Acta Math.*
32. O. Kellogg, *Foundations of Potential theory*, Springer-Verlag, 1929 (reprinted by Dover Books, 1954).
33. C. Kenig, "Weighted H^p spaces on Lipschitz domains," *Amer. J. Math.*, vol. **102** (1980), 129–163.
34. ———, "Weighted Hardy spaces on Lipschitz domains," *Proc. of Symposia in Pure Math.*, vol. **35**, part 1 (1979), 263–274.
35. J. Lions and E. Magenes, *Problèmes aux limites non Homogènes, et Applications*, t. I, Dunod, 1968.
36. J. Marcinkiewicz and A. Zygmund, "A Theorem of Lusin," *Duke Math. J.*, **4** (1938), 473–485.
37. C. Morrey Jr., *Multiple Integrals in the Calculus of Variations*, Springer-Verlag, NY, 1966.

38. B. Muckenhoupt, "Weighted norm inequalities for the Hardy maximal function," *Trans. Amer. Math. Soc.*, **165** (1972), 207–226.

39. ———, "The equivalence of two conditions for weight functions," *Studia Math.*, **49** (1974), 101–106.

40. L. Payne and H. Weinberger, "New bounds in harmonic and biharmonic problems," *J. Math. Phys. Sci.*, **33** (1954), 291–307.

41. F. Rellich, "Darstellung der Eigenwerte von $\Delta u + \lambda u$ durch ein Randintegral," *Math. Z.*, **46** (1940), 635–646.

42. D. Spencer, "A function theoretic identity," *Amer. J. Math.*, **65** (1943), 147–160.

43. E. M. Stein, "On the theory of harmonic functions of several variables II, Behavior near the boundary," *Acta Math.*, **106** (1961), 137–174.

44. ———, *Singular Integrals and Differentiability Properties of Functions*, Princeton University Press, Princeton, NJ, 1970.

45. E. M. Stein and G. Weiss, "On the theory of harmonic functions of several variables I," *Acta Math.*, **103** (1960), 25–62.

46. N. Wiener, "The Dirichlet problem," *J. Math. Phys. Sci.*, **3** (1924), 127–146.

February 1981

MINIMAL SURFACES AND PARTIAL DIFFERENTIAL EQUATIONS

Johannes C. C. Nitsche

This article is divided into two parts. The first part contains a brief exposition of the historical development of the theory of minimal surfaces. It is neither meant to be, nor can it be exhaustive; but in alluding to a number of interesting discoveries from early and new times and to special problems, it will give the reader a general impression of the beauty and riches with which this area of mathematics is endowed. The first part might be entitled, "The story of minimal surfaces." The second part is more technical. It presents proofs and describes in detail some important properties of minimal hypersurfaces — gradient estimates, Bernstein's theorem, removable and non-removable singularities — viewed from the vantage point of partial differential equations, as well as the theorem on the boundary regularity of solutions of Plateau's problem. The literature on the subject is vast, and it seemed advisable to keep bibliographical attributions to a minimum. Extensive references to the matters discussed here can be found in the works listed at the end. A few of the new results mentioned have not yet appeared in print.

PART 1

The history of minimal surfaces begins with J. L. Lagrange whose famous memoir of 1762 includes the following example for the new variational calculus devised by him: Suppose Γ is a closed curve in Euclidean space; to find a surface of least area bounded by Γ. If Γ has a simply-covered projection onto a plane, say the (x, y)-plane, and if D is the interior of this projection, we may assume that the surface $\mathbb{S}$ in question can be represented over D in the form $z = f(x, y)$. Because the area

$$A[\mathbb{S}] = \iint_D \sqrt{1 + f_x^2 + f_y^2} \, dx \, dy$$

is to be a minimum, it cannot be diminished when $f(x, y)$ is replaced by $f(x, y) + t\zeta(x, y)$, where t is a small number and $\zeta(x, y)$ is an arbitrary function which vanishes on ∂D (so that the comparison surface $\mathbb{S}(t): z = f(x, y) + t\zeta(x, y)$ is again bounded by Γ). Thus,

$$\frac{d}{dt} A[\mathbb{S}(t)]\Big|_{t=0} = 0$$

and a familiar argument based on the discretion in the choice of $\zeta(x, y)$ shows that the relation

$$\frac{\partial}{\partial x}\left(\frac{f_x}{\sqrt{1 + f_x^2 + f_y^2}}\right) + \frac{\partial}{\partial y}\left(\frac{f_y}{\sqrt{1 + f_x^2 + f_y^2}}\right) = 0 \qquad (1.1)$$

or equivalently,

$$\mathfrak{L}[f] \equiv \left(1 + f_y^2\right)f_{xx} - 2f_x f_y f_{xy} + \left(1 + f_x^2\right)f_{yy} = 0 \qquad (1.2)$$

must be satisfied as a necessary condition for the solution surface $\mathbb{S}$. (1.1) can be interpreted as integrability condition for the differential forms

$$p\,dx + q\,dy \quad \text{and} \quad \frac{q\,dx - p\,dy}{\sqrt{1 + p^2 + q^2}},$$

where $p = f_x(x, y)$, $q = f_y(x, y)$. Lagrange states, "It is clear that these conditions can be satisfied if p and q are assumed to be constants, a choice which leads to a plane as the required surface. This, however, is only a very special case, because the general solution must have the property that the boundary of the surface can be assumed to have an arbitrary shape."

Lagrange himself did not pursue the matter further. It was soon discovered that the catenoid $z = \cosh^{-1}\sqrt{x^2 + y^2}$ and the helicoid $z = \tan^{-1}(y/x)$ satisfy his conditions. In 1776 J. B. M. C. Meusnier provided a geometrical interpretation: The left-hand side in (1.1) is twice the mean curvature of $\mathbb{S}$. (The mean curvature H of a surface at a given point is the average of the principal curvatures at this point.) Because of their provenance, surfaces of vanishing mean curvature have been called minimal surfaces ever since, despite the fact that such surfaces often do not realize a minimum (not even a relative minimum) of area. The question whether they do depends on the second variation

$$\frac{d^2}{dt^2}A[\mathbb{S}(t)]\bigg|_{t=0}$$

and leads to the eigenvalue problem for the differential equation $\Delta_\mathbb{S}\eta + \lambda 2|K|\eta = 0$. Here $\Delta_\mathbb{S}$ denotes the Beltrami operator, and K is the Gaussian curvature of $\mathbb{S}$.

G. Monge (1784), A. Legendre (1787), and later S. F. Lacroix, A. M. Ampère, and others made first attempts at the integration of (1.2), now called the minimal surface equation. Fifty years passed before further explicit examples of minimal surfaces were disclosed by H. F. Scherk. Of the five new surfaces we mention here the surface

$$z = \log\cos y - \log\cos x, \tag{1.3}$$

which is real only over the "white squares" $|x - (m + n)\pi| < \pi/2$, $|y - (m - n)\pi| < \pi/2$ of an infinite chessboard, and the surface

$$\sin z = \sinh x \sinh y. \qquad (1.4)$$

During the middle of the last century, the uncovering of new minimal surfaces, and facts about them, had taken a rapid pace. To the mathematician who can derive pleasure from his intimate acquaintance with a geometric object—and there were many: E. Catalan, O. Bonnet, J. A. Serret, B. Riemann, K. Weierstrass, A. Enneper, H. A. Schwarz, J. Weingarten, L. Lindelöf, E. Beltrami, U. Dini, A. Ribaucour, E. R. Neovius, L. Bianchi, S. Lie, A. Cayley, A. Schoenflies, etc.—each revealed its special treasures. It was one of Lie's most beautiful discoveries (and perceived by him as such) that Scherk's surface (1.3) can be generated in infinitely many ways as a translation surface. It shares this property with the helicoids. There is a curious connection with Abel's theorem for algebraic curves of fourth order. Lie's investigations of the surfaces which can be generated in more than one way as translation surfaces were generalized in 1938 by W. Wirtinger to higher dimensional manifolds.

It might be a challenge for the reader to determine all minimal surfaces with a representation of the special form $f(x) + g(y) + h(z) = 0$. (This determination has been carried out by J. Weingarten in 1887 and again by M. Fréchet in 1956–57.) Obviously, catenoid, helicoid, as well as the surfaces (1.3) and (1.4) belong to this category; but there are many others, including the minimal surface with the representation

$$\mathscr{E}(x)\mathscr{E}(y)\mathscr{E}(z) = 1, \qquad (1.5)$$

where $t = \mathscr{E}(\xi)$ denotes the inverse function of the elliptic integral

$$\xi = \int_0^t \frac{d\tau}{\sqrt{1 + \tau^2 + \tau^4}}. \qquad (1.5')$$

This function is a solution of the differential equation

$$\mathscr{E}'^2 = a\mathscr{E}^4 + b\mathscr{E}^2 + c \qquad (1.5'')$$

for the choice of the constants $a = b = c = 1$. (The determination of all minimal surfaces (1.5), (1.5″) was accomplished in 1887 by A. Cayley.) The part of (1.5) within the cube

$$0 \leqslant x, y, z \leqslant \xi_0 = \int_0^\infty \frac{d\tau}{\sqrt{1 + \tau^2 + \tau^4}} = \frac{2}{3} K\left(\frac{2}{3}\sqrt{2}\right) = 1.68575$$

is a simply-connected surface portion $\mathfrak{S}$ which is bounded by six edges of this cube forming a skew hexagon, as depicted in Figure 1. The surface contains three straight lines (bold in Figure 1) through the center of the cube, connecting the midpoints of opposite sides. This follows from the fact that, for $0 < t < \infty$,

$$\int_0^t \frac{d\tau}{\sqrt{1 + \tau^2 + \tau^4}} = \int_{1/t}^\infty \frac{d\tau}{\sqrt{1 + \tau^2 + \tau^4}},$$

so that

$$\mathfrak{S}(\xi_0/2) = \int_0^1 \frac{d\tau}{\sqrt{1 + \tau^2 + \tau^4}},$$

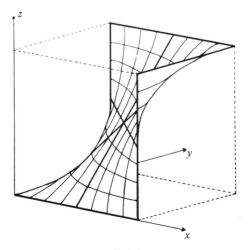

Fig. 1

74   *Johannes C. C. Nitsche*

and that $\mathfrak{S}(\xi)\mathfrak{S}(\eta)=1$ implies $\xi+\eta=\xi_0$. $\mathfrak{S}$ can thus be decomposed into six congruent minimal surfaces spanned into congruent skew quadrilaterals.

Historically speaking, the first successful solution of the problem posed by Lagrange for a nontrivial contour, *ab ovo* —"first the contour, then the surface"—was established in 1865 by Schwarz and also by Riemann. This surface is shown in Figure 2. Its (parametric) representation, with the help of a complex variable $w = u + iv$, is

$$x = \operatorname{Re} \int^{w}(1-\omega^2)R(\omega)\,d\omega, \quad y = \operatorname{Re} \int^{w}i(1+\omega^2)R(\omega)\,d\omega,$$

$$z = \operatorname{Re} \int^{w}2\omega R(\omega)\,d\omega, \tag{1.6}$$

where

$$R(\omega) = \frac{1}{\sqrt{1-14\omega^4+\omega^8}}. \tag{1.6'}$$

The function $R(\omega)$ has singularities at the eight points $\omega = \pm(\sqrt{3}\pm1)/\sqrt{2}, \pm i(\sqrt{3}\pm1)/\sqrt{2}$. It is interesting to note that the

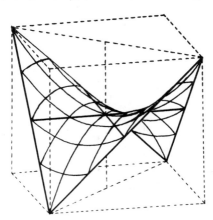

Fig. 2

detailed study of the three hyperelliptic integrals in (1.6) and their periodicity properties occupies much of Schwarz's *Collected Works*. (Equally impressive is the extent of Lie's opus dedicated to minimal surfaces.) Because a minimal surface can be extended as minimal surface by reflection across straight segments of its boundary and because the edges of a regular tetrahedron meet at 60°-angles, it is possible to use the Schwarz-Riemann surface as a building block (fundamental domain) for larger surfaces. Figures 3a and 3b show the composition of six congruent specimens to a disc-type and a ring-type minimal surface, respectively. Both surfaces are inscribed in a cube. Interesting formations are obtained by further extension. A surface composed of 18 original minimal surfaces is depicted in Figure 4. It is orientable, of genus $g = 0$ (Euler characteristic $\chi \equiv 2 - 2g - r = -2$), and its $r = 4$ borders lie on the sides of a cube, and also on the sides of a tetrahedron. Unlimited continuation leads to a complete minimal surface embedded in space. This periodic minimal surface of infinite genus contains infinitely many straight lines; it also effects a peculiar division of space into two disjoint, but highly linked domains of infinite connectivity. Recently, G. Donnay and D. L. Pawson have made the interesting observation that the interface between inorganic crystalline and organic amorphous matter in the skeletal elements of echinoderms have a striking similarity to such periodic minimal surfaces. The

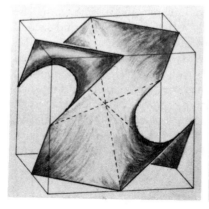

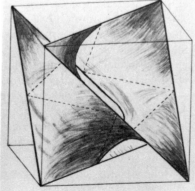

Fig. 3a   Fig. 3b

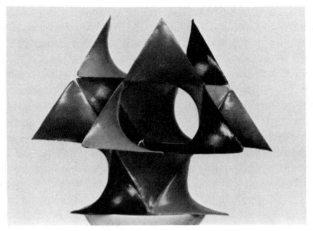

Fig. 4*

work of these authors, as well as H. L. Nissen, contains convincing pictures (*Science*, **166** (1969), 47–52).

The visual impression that the minimal surface of Figure 4 contains circles and "catenoidal" parts will not bear closer examination. There are, however, periodic cyclic minimal surfaces. Enneper proved in 1869 that if a minimal surface is generated by a one-parameter family of circles, the circles must lie in parallel planes. The explicit determination of these surfaces, in terms of elliptic integrals, was carried out by Enneper and by Riemann. The catenoid is the simplest example. By estimating the first and second characteristic values of an associated eigenvalue problem, M. Shiffman showed in 1956 that a ring-type minimal surface bounded by two circles in parallel planes must be cyclic. The question whether two circles in parallel planes are capable of bounding such a surface leads to a delicate determination of constants. There can be never more than two solution surfaces, one furnishing a relative minimum of area, the other unstable. If the circles are too far apart, or if their projections (orthogonal to their planes) do not overlap, then no solution surface can exist (J. C. C. Nitsche, 1968).

Helicoids are generated by straight lines, cyclic minimal surfaces by circles. One might ask for minimal surfaces to be generated by

---

*Figure 4 gives another view of the model shown in [25], p. 246.

1-parameter families of other specific curve classes. It is known that minimal surfaces, other than planes, cannot be generated by hyperbolas or (noncircular) ellipses. On the other hand, E. Catalan (1855) and A. Enneper (1882) have found minimal surfaces which contain families of parabolas.

Minimal surfaces have clear-cut intersection properties: If two minimal surfaces $\mathbb{S}_1$ and $\mathbb{S}_2$ have one point in common, their intersection locus consists locally of finitely many analytic arcs $\gamma_1, \gamma_2, \ldots, \gamma_n$ forming equal angles $\angle(\gamma_j, \gamma_{j+1}) = \pi/n$ at this point. In particular, $\mathbb{S}_2$ cannot stay entirely on one side of $\mathbb{S}_1$. For the proof, represent the surfaces locally in the form $z = z^{(k)}(x, y)$ satisfying

$$z^{(k)}(0,0) = z_x^{(k)}(0,0) = z_y^{(k)}(0,0) = 0,$$

so that

$$z^{(k)}(x, y) = \sum_{m = m_k}^{\infty} P_m^{(k)}(x, y),$$

where the $P_m^{(k)}(x, y)$ are homogeneous polynomials of degree m and $m_k \geqslant 2$. Substitution into equation (1.2) shows that the first term in the expansion of the difference $z^{(2)}(x, y) - z^{(1)}(x, y)$ is a harmonic polynomial.

Physical realizations of minimal surfaces can be obtained through soap films spanned into a frame fashioned of thin wire. Such films are extremely thin so that the influence of gravitation can be neglected and, although they are actually bounded by two skins, they present the image of ideal surfaces. Since the latter are generally open, that is, enclose no volume, both of their sides are subject to the same pressure and their mean curvature must be zero everywhere. This follows from the fact that the normal pressure on an interface, due to the action of surface tension, is, per unit area,

$$p = \sigma(\kappa_1 + \kappa_2) = 2H. \tag{1.7}$$

Here σ is the surface (or interfacial) tension and κ_1, κ_2 denote the principal curvatures. Closed soap films—bubbles—are surfaces of constant, but nonzero mean curvature. For other capillary phenomena, the mean curvature of the interface separating two media may well be a variable function. Such an interface $z = z(x, y)$ satisfies the Laplace-Young equation

$$\frac{\partial}{\partial x}\left(\frac{z_x}{\sqrt{1 + z_x^2 + z_y^2}}\right) + \frac{\partial}{\partial y}\left(\frac{z_y}{\sqrt{1 + z_x^2 + z_y^2}}\right) = 2H(x, y, z).$$

(1.8)

For the boundary of a liquid drop, with the gravitational forces acting in z-direction, one has $H = az + b$. If the drop is free from the influence of gravity—by being suspended in another immiscible liquid of equal density or in a vehicle in outer space—then $H = $ const. In this case the drop assumes spherical shape, unless it rotates or is otherwise subject to external forces, for example the influence of an electric field.

In honor of J. Plateau (1801–1883) and his fundamental experiments illuminating the effects of surface tension, the problem of finding a minimal surface spanned into a given contour has become known as *Plateau's problem*. The charm of these experiments, which often produce the most bizarre surface formations, has been acclaimed many times. In his obituary notice for Plateau, G. Van der Mensbrugghe writes: "There is, to our knowledge, no example where experimental observation has supported the theory with more ravishing forms. What could appear more beautiful to the eyes of a mathematician than these weightless models, adorned with the most brilliant colors, and, despite their extraordinary fragility, of an astonishing persistence." It is a tragic biographic footnote that Plateau, whose scientific papers on the subject appeared during the period 1843–1869, could not view his own experiments; he had turned totally blind in 1843. In the 1980's, the traditional soap film experiments are being replaced by computer graphics developed in various quarters.

Plateau did not only describe how a minimal surface defined by mathematical equations could be realized experimentally, he also formulated the distillate of his observations as a postulate: that an arbitrarily prescribed contour always bounds a minimal surface which adheres to it in its full extent. The attempts to provide a rigorous existence proof in support of the physical evidence posed—and still poses today—a major challenge for mathematicians.

In the beginning, after successful determination of the surface (1.6), (1.6′) through a special skew quadrilateral, Plateau's problem was attacked for a number of simple polygonal contours. The solution program was outlined by Weierstrass and later, in greater detail, by Darboux. A brief description follows. The spherical mapping of a minimal surface $\mathfrak{S}$ is anticonformal, albeit not injective near the umbilical points of $\mathfrak{S}$. (This is a characteristic property of minimal surfaces.) A subsequent stereographic projection and, possibly, further mappings lead to a conformal relation between $\mathfrak{S}$ and a domain in a complex $w = u + iv$-plane or a Riemann surface. The components x, y, z of the position vector of $\mathfrak{S}$ become functions of u, v, and the line element of $\mathfrak{S}$ assumes the form $ds^2 = E(u, v)(du^2 + dv^2)$. For the isothermal parameters u, v the vanishing of the mean curvature of $\mathfrak{S}$ is equivalent to the condition that $x(u, v), y(u, v), z(u, v)$ be harmonic functions. Thus there are three analytic functions $g_1(w), g_2(w), g_3(w)$ such that

$$x = \operatorname{Re} g_1(w), \quad y = \operatorname{Re} g_2(w), \quad z = \operatorname{Re} g_3(w). \qquad (1.9)$$

These functions must satisfy Monge's relation

$$g_1'^2(w) + g_2'^2(w) + g_3'^2(w) = 0. \qquad (1.10)$$

(1.10) can be parametrized with the help of two independent analytic functions. Various such parametrizations have been proposed. The best known, and geometrically most meaningful, is that of Weierstrass:

$$g_1'(w) = \Phi^2(w) - \psi^2(w), \quad g_2'(w) = i\left[\Phi^2(w) + \psi^2(w)\right],$$

$$g_3'(w) = 2\Phi(w)\psi(w), \qquad (1.11)$$

where $\Phi(w)$ and $\psi(w)$ have no common zero. From the expression for the unit normal vector of $\mathfrak{S}$, which follows from (1.11),

$$\mathfrak{X} = \left\{ \frac{\Phi\bar\psi + \bar\Phi\psi}{|\Phi|^2 + |\psi|^2}, \frac{i(\Phi\bar\psi - \bar\Phi\psi)}{|\Phi|^2 + |\psi|^2}, \frac{|\psi|^2 - |\Phi|^2}{|\Phi|^2 + |\psi|^2} \right\}$$

$$= \left\{ \frac{\omega + \bar\omega}{1 + |\omega|^2}, \frac{i(\bar\omega - \omega)}{1 + |\omega|^2}, \frac{|\omega|^2 - 1}{1 + |\omega|^2} \right\}, \tag{1.12}$$

the geometrical significance of the quantity $\omega = \psi/\Phi$ becomes evident: It is the complex variable in the equatorial plane of the unit sphere obtained from the spherical image of $\mathfrak{S}$ by stereographic projection. We also observe that (1.6) is a special case of the representation (1.9), (1.11) for surface portions free of umbilical points. In this way, the catenoid is obtained for the special choice $R(\omega) = 1/2\omega^2$, the helicoid for the choice $R(\omega) = i/2\omega^2$, Scherk's surface (1.3) for the choice $R(\omega) = 2/(1 - \omega^4)$, etc.

A rigid motion of $\mathfrak{S}$ leads to the representation (1.9), (1.11) of $\mathfrak{S}$ with the help of new analytic functions $\tilde\Phi(w)$, $\tilde\psi(w)$ which are obtained from the functions $\Phi(w)$ and $\psi(w)$ by a Cayley transformation:

$$\tilde\Phi(\omega) = a\Phi(\omega) + b\psi(\omega), \quad \tilde\psi(w) = -\bar b\Phi(w) + \bar a\psi(w),$$

$$|a|^2 + |b|^2 = 1. \tag{1.13}$$

For instance, a clockwise rotation of $\mathfrak{S}$ about the z-axis by an angle λ is effected with the choice $a = e^{i\lambda/2}$, $b = 0$. Both functions $\Phi(w)$ and $\psi(w)$ are solutions of the ordinary differential equation of second order,

$$\Xi''(w) + p(w)\Xi'(w) + q(w)\Xi(w) = 0 \tag{1.14}$$

with the coefficients

$$p(w) = -\frac{\Phi\psi'' - \psi\Phi''}{\Phi\psi' - \psi\Phi'}, \quad q(w) = -\frac{\Phi\psi'' - \psi\Phi''}{\Phi\psi' - \psi\Phi'}. \tag{1.15}$$

These coefficients are invariant under all transformations (1.13). It is also easily seen that they are real on any straight line segment contained on $\mathcal{S}$.

Now let Γ be a polygon in space with vertices $\mathfrak{p}_1, \mathfrak{p}_2, \ldots, \mathfrak{p}_m$ and $\mathcal{S}$ a minimal surface bounded by Γ. Assume that $\mathcal{S}$ is the image of the unit disc $|w| \leqslant 1$ in the complex w-plane and that the vertices $\mathfrak{p}_k$ correspond to the points $w_k = e^{i\theta_k}$ on $|w| = 1$. Of course, the exact value of the angles θ_k is unknown at the start; their determination is one of the goals of the solution process. The functions $\Phi(w)$ and $\psi(w)$ are analytic in $|w| < 1$ and can be continued analytically across each open arc of $|w| = 1$ between the points w_k, but they will have singularities at the points w_k. As the arc $w = e^{i\theta}$, $\theta_k < \theta < \theta_{k+1}$ is crossed, the image in space progresses from the surface $\mathcal{S}$ to its extension obtained by reflection on the side $\overline{\mathfrak{p}_k \mathfrak{p}_{k+1}}$ of Γ. If we let the variable w go once around the point w_k on a small closed curve, its image moves from the surface $\mathcal{S}$ to its reflection on the side $\overline{\mathfrak{p}_{k-1} \mathfrak{p}_k}$ of Γ and on to the reflection of the latter on the side $\overline{\mathfrak{p}_k \mathfrak{p}_{k+1}}$. The concomitant analytic continuation of the functions $\Phi(w)$ and $\psi(w)$ leads to new functions $\tilde{\Phi}(w)$, $\tilde{\psi}(w)$. Because the two successive reflections amount to a rotation about the line orthogonal to the segments $\overline{\mathfrak{p}_{k-1} \mathfrak{p}_k}$ and $\overline{\mathfrak{p}_k \mathfrak{p}_{k+1}}$ (by an angle which is twice the angle included between these segments), $\tilde{\Phi}(w)$ and $\tilde{\psi}(w)$ are obtained from $\Phi(w)$ and $\psi(w)$ by a Cayley transformation (1.13) with coefficients a, b which depend on the geometric properties of Γ near the vertex $\mathfrak{p}_k$. In view of what has been said earlier, the functions $p(w)$ and $q(w)$ in (1.14), (1.15) are single-valued in the whole w-plane with isolated singularities at the points w_k and possibly at certain other points, where the denominator $\Phi\psi' - \psi\Phi'$ vanishes. Each point w_k is associated in this way with a Cayley transformation dependent on the directions of successive sides of Γ. The solution of Plateau's problem is now seen to be related to the problem of determining the functions $\Phi(w)$, $\psi(w)$ which are analytic in the extended w-plane except at the points $w_1, w_2, \ldots, w_m$ where they may have specified singularities and which undergo prescribed linear transformations along circuits around these points. This problem, conceived by Riemann in connection with his studies of the global theory of linear differential equations and brought to the attention of mathematicians again by Hilbert in his famous Paris lecture of 1900 as the 21st problem, has been investigated by many authors, among them L. Schlesinger, J. Plemelj, G. D. Birkhoff, R. Garnier, and in recent years N. I. Muskhelishvili, K. Okubo, H. Röhrl, et al. The problem at hand, of course, has additional difficulties: The location of the points w_k on $|w| = 1$ must be determined in such a way—there may be several possibilities—that each arc $\widehat{w_k w_{k+1}}$ is mapped precisely onto the corresponding side $\overline{\mathfrak{p}_k \mathfrak{p}_{k+1}}$ of Γ. This determination is a formidable task; it poses difficulties even for rather simple contours, and it requires a careful study of the local and global properties of the coefficients $p(w)$ and $q(w)$ in the differential equation (1.14). From the properties of $\Phi(w)$ and $\psi(w)$ and from (1.15), it can be concluded that the points w_k are regular singular points, in the sense of Fuchs, for the differential equation (1.14).

The solution program sketched above was carried to a successful completion by R. Garnier in 1928. An approximation procedure leads also to the solution of Plateau's problem for more general contours. Due to the complexities involved, Garnier's approach is not any more well understood these days, if it ever was. This is deplorable because it appears that a continued study of the dependence of the points w_k on the direction and length of the sides $\overline{\mathfrak{p}_k \mathfrak{p}_{k+1}}$ of Γ might provide the answer to certain questions regarding the solutions of Plateau's problem—at least for the class of polygonal boundaries—which are still open today. It is for instance unknown whether a contour can bound a continuum of minimal surfaces, somewhat in the way in which antipodal points on the globe are connected by a continuum of geodesics (great circles). In 1979, E. Heinz could draw on Schlesinger's results to prove the analytic dependence of the solution vector on the $\mathfrak{p}_k$ for a boundary value problem related to Plateau's problem. In this problem, which was conceived by Shiffman, the image of each arc $\widehat{w_k w_{k+1}}$ is merely required to lie on the straight line through the vertices $\mathfrak{p}_k$ and $\mathfrak{p}_{k+1}$, but may "overshoot" these points.

Shortly after Garnier's work had appeared, both J. Douglas and T. Radó published their famous approaches to Plateau's problem. The power of the direct methods of the calculus of variations was prominently displayed by these solutions. Later simplifications suggested by R. Courant and L. Tonelli have brought the existence proof within reach for an advanced calculus course. There were parallel efforts, based in part on earlier ideas of Lebesgue, by E. J. McShane, C. B. Morrey and others, and later extensions of the problem to minimal surfaces in Riemannian manifolds by A. Lonseth (1942) and especially C. B. Morrey (1948). Continued attention has also been devoted to Plateau's problem in its non-parametric form—that is: to the Dirichlet problem for the minimal surface equation (1.2), (1.24)—from the work of S. Bernstein, A. Korn, C. H. Müntz, and A. Haar to the deep new investigations on nonuniformly elliptic differential equations and systems.

The Douglas-Radó solution is a minimal surface $\mathcal{S}$ parametrized over the closure of the unit disc

$$P = \{u, v; \; u^2 + v^2 < 1\}$$

in the form (1.9), (1.10). Its position vector $\mathfrak{x}(u, v) = \{x(u, v), y(u, v), z(u, v)\}$ has the following properties:

(i) $\mathfrak{x}(u, v) \in C^2(P) \cap C(\bar{P})$;

(ii) $\Delta \mathfrak{x} = 0$, $\mathfrak{x}_u^2 = \mathfrak{x}_v^2$, $\mathfrak{x}_u \mathfrak{x}_v = 0$ in P;

(iii) $\mathfrak{x}$ provides a bijective mapping between ∂P and Γ such that three prescribed points on ∂P correspond to three prescribed points on Γ.

We shall say that the surface $\mathcal{S} = \{\mathfrak{x} = \mathfrak{x}(u, v); \; (u, v) \in \bar{P}\}$ is a solution of Plateau's problem for the contour Γ if the position vector $\mathfrak{x}(u, v)$ obeys the above conditions.

Property (ii) insures that the area of $\mathcal{S}$ can be expressed by Dirichlet's integral

$$D[\mathfrak{x}] = \frac{1}{2} \int\int_P \left(\mathfrak{x}_u^2 + \mathfrak{x}_v^2 \right) du \, dv;$$

in fact, the solution process is based on the minimization of this integral. If Γ is rectifiable (and thus capable of bounding disc-type surfaces of finite area), the Douglas-Radó solution $\mathcal{S}$ is a surface of least area; otherwise, $\mathcal{S}$ is obtained with the help of an approximation argument.

The occurrence of isolated zeros for the gradient of $\mathfrak{x}(u, v)$ could not be excluded by Douglas and Radó. Near such points—they are called branch points—the mapping by the vector $\mathfrak{x}(u, v)$ is no immersion, and $\mathfrak{S}$ loses its differential geometric character. Depending on the nature of the local expansion for $\mathfrak{x}(u, v)$, a branch point may be "true" or "false." False branch points indicate a poor choice of parameters; they can be removed through local reparametrizations, but not in general by global reparametrizations (without violating property (iii)).

There are certain geometric conditions guaranteeing the absence of branch points a priori. For example, if there is a line l in space such that every plane through l intersects Γ in at most two points, then no solution of Plateau's problem for Γ has interior branch points. (This condition is satisfied if Γ has a simply-covered projection onto a plane starshaped curve.) For the general case, however, it took until 1970, when an argument was put forth by R. Osserman which showed that minimal surfaces of least area cannot carry interior branch points. Subsequently, R. D. Gulliver and F. D. Lesley proved the same theorem for boundary branch points, provided that the contour Γ is known to be analytic. Whether minimal surfaces of least area spanned into a regular, but nonanalytic, contour can possess boundary branch points is unknown today. We remark that unstable minimal surfaces may contain interior as well as boundary branch points.

Another desideratum regarding the Douglas-Radó solution stemmed from the lack of information about the boundary behavior of $\mathfrak{S}$. It was known from the beginning that a solution of Plateau's problem is analytic on straight parts of the bounding contour Γ and that it can be extended (as a minimal surface) by reflection across such segments. H. Lewy proved in 1951 that $\mathfrak{x}(u, v)$ is analytic on each arc of ∂P which corresponds to an analytic subarc of Γ. More recently, through the efforts of several mathematicians (E. Heinz, S. Hildebrandt, D. Kinderlehrer, F. D. Lesley, J. C. C. Nitsche, F. Tomi, and S. Warschawski) the boundary behavior of $\mathfrak{x}(u, v)$ has become fully understood. In particular, the following theorem reminiscent of Kellogg's theorem in the theory of conformal mapping holds (Hildebrandt: $m \geqslant 4$, Nitsche: $m \geqslant 1$; 1969):

If Γ is a regular Jordan curve of class $C^{m,\alpha}$ ($m \geqslant 1$, $0 < \alpha < 1$), then $\mathfrak{x}(u, v)$ belongs to the regularity class $C^{m,\alpha}(\bar{P})$. The norm

$\| \mathfrak{x} \|_{m, \alpha}^{\bar{P}}$ *can be estimated uniformly for all solutions of Plateau's problem.*

The behavior of $\mathfrak{S}$ near a corner of its boundary formed by two subarcs of Γ has been elucidated by G. Dziuk (1981).

The Kellogg-type theorem above will alert the reader to the fact that Plateau's problem may have several, even infinitely many solutions. There are many simple examples illustrating this phenomenon. Consider for instance the part

$$\mathfrak{S}_{r, \alpha} = \{ x = u \cos v, y = u \sin v, z = v; \ |u| \leqslant r, \ |v| \leqslant \alpha \pi \}$$

of the helicoid $\tan z = y/x$ which is spanned into a curve $\Gamma_{r, \alpha}$ composed of two congruent helical parts and two segments. $\Gamma_{r, \alpha}$ bounds at least two solutions of Plateau's problem if $\alpha > 1/2$ and $r > \bar{r}(\alpha)$. Here $\bar{r}(\alpha)$ denotes the root of the equation

$$\bar{r}(\alpha) = \sinh \bar{s}, \ \frac{1}{2\alpha} = \tanh(\bar{s}) \tanh\left(\frac{\bar{s}}{2\alpha} \right).$$

(We have $\bar{r}(3/4) = 2$, $\bar{r}(1) = \sqrt{3}$, $\bar{r}(3/2) = [(1 + \sqrt{3})/2]^{3/2}$, etc.)

As had been mentioned at the beginning, the question of stability—whether the area of a minimal surface $\mathfrak{S} = \{ \mathfrak{x} = \mathfrak{x}(u, v); \ (u, v) \in \bar{P} \}$ is smaller than that of all neighboring surfaces with the same boundary—is intimately related to the eigenvalue problem

$$\Delta_{\mathfrak{S}} \eta + \lambda 2 |K| \eta = 0 \quad \text{in} \quad P, \quad \eta = 0 \quad \text{on} \quad \partial P.$$

$\mathfrak{S}$ is stable if the first eigenvalue λ_1 is greater than one. It was in the context of minimal surfaces that H. A. Schwarz developed his fundamental treatment of eigenvalue problems. He also discovered beautiful geometrical interpretations: The surface $\mathfrak{S}$ is stable if its spherical image is entirely contained in a hemisphere (and thus has area $< 2\pi$) or if its tangent planes do not meet a specific point in space. The second criterion implies, among other things, L. Lindelöf's well-known construction of conjugate points on a catenary. The first criterion has been generalized in an essential way by J. L. Barbosa and M. do Carmo (1976): The surface $\mathfrak{S}$ is stable whenever the area of its spherical image is smaller than 2π.

Some of the solutions of Plateau's problem for a given contour Γ may be unstable, as is the helicoidal portion $\mathfrak{S}_{r,\alpha}$ in the example. It is clear that the existence proof for unstable minimal surfaces is beyond the scope of any solution procedure involving minimization techniques. Rather, new approaches, notably approaches utilizing the methods of Morse theory, have to be employed. Here we have the ground-breaking investigations of 1939 by Morse-Tompkins and by Shiffman. The fact that the area functional is not continuous on the underlying space of contours causes considerable difficulties. A representative theorem is the following:

If the curve Γ (assumed to satisfy mild regularity conditions) bounds two solutions of Plateau's problem, the position vector of each providing a strict local minimum for Dirichlet's integral, then there exists a third, unstable solution of Plateau's problem for Γ.

The proof utilizes an argument going back to G. D. Birkhoff and nowadays described in graphic terms as the mountain pass lemma. A modified approach to the theorem is due to R. Courant (1941) who investigates first simple polygonal contours for which the problem is one of finitely many dimensions. Generalizations to unstable minimal surfaces in Riemannian spaces have been obtained by G. Ströhmer (1980).

To date, our knowledge concerning the totality of solutions of Plateau's problem is limited. The ideal result—at the present time out of reach—would be a statement about the number of solutions, in terms of the geometric properties of Γ. Such precise information is now available only in the few cases for which uniqueness is assured beforehand:

(a) If Γ is a plane curve. Here Plateau's problem is reduced to Riemann's mapping theorem of complex function theory.

(b) If Γ has a simply-covered parallel or central projection onto a plane curve (Radó, 1932).

(c) If Γ is analytic and of total curvature $\kappa(\Gamma) \leqslant 4\pi$ (Nitsche, 1973). If $\kappa(\Gamma) < 4\pi$, it suffices to assume that Γ belong to the regularity class $C^{2,\alpha}$.

The proof of statement (c) employs, among other things, an interesting observation of S. Sasaki (1959) relating the total curvature of Γ to the number of possible branch points on $\mathfrak{S}$. In particular, $\mathfrak{S}$ cannot carry branch points at all if $\kappa(\Gamma) < 4\pi$. Sasaki's inequality is based on the Gauss-Bonnet theorem and can be used to prove W. Fenchel's theorem which states that the total curvature of a closed space curve cannot be smaller than 2π.

If an analytic contour Γ is known to bound only minimal surfaces free of branch points, then the inequality $\kappa(\Gamma) \leqslant 6\pi$ insures the existence of at most finitely many solutions of Plateau's problem (Nitsche, 1978). For an analytic contour Γ, also the number of Douglas-Radó solutions (of least area) is known to be finite (Tomi, 1973). This finiteness property has been established, at least generically, to be of general validity by the new abstract theory which is being developed by A. Tromba and R. Böhme: There exists an open dense subset in the space of curves with the C^∞ topology, such that each contour of this subset bounds only finitely many solutions of Plateau's problem. A similar statement regarding the absolute minima is due to F. Morgan, 1978. Morgan's theorem substantiates the heuristic expectation that in cases where there are several surfaces of least area, a small deformation of the boundary will separate them, restoring uniqueness for the absolute minimum.

These generic statements contain no information about the contours Γ in the exceptional set, although this set might be quite large. It is plausible that a "reasonable" contour cannot bound infinitely many solutions of Plateau's problem. But are there any contours which bound infinitely many, or even continua of disc-type minimal surfaces? That such contours might exist has been suggested, heuristically, by P. Lévy and by R. Courant thirty years ago. The arguments can be based on the so-called bridge theorem which may, somewhat loosely, be stated as follows:

Let Γ_1 and Γ_2 be Jordan curves bounding two solutions $\mathfrak{S}_1$ and $\mathfrak{S}_2$ of Plateau's problem, respectively. Form a new contour Γ by removing a small arc on each of these curves and connecting the newly created end points by a "bridge" of two nearly parallel arcs of bounded length at a *distance not exceeding the value* $\varepsilon > 0$. *Then* Γ *will bound a solution of Plateau's problem consisting of three parts* $\mathbb{S}^{(0)}$, $\mathbb{S}^{(1)}$ *and* $\mathbb{S}^{(2)}$. *With* ε *tending to zero,* $\mathbb{S}^{(1)}$ *and* $\mathbb{S}^{(2)}$ *converge to* $\mathbb{S}^{(1)}$ *and* $\mathbb{S}^{(2)}$, *respectively, while the area of* $\mathbb{S}^{(0)}$ *spanning the bridge converges to zero.*

Let us turn our attention to a Jordan curve which is capable of bounding at least two solutions of Plateau's problem of different areas, A and $B > A$. In the configuration to be constructed presently, this curve will be taken as our first curve Γ_1. Next to Γ_1, at a distance d from it, a curve Γ_2 is placed, similar to Γ_1 in shape but $1/\sqrt{2}$ times smaller. Next to Γ_2, at distance $d/\sqrt{2}$ from it, a curve Γ_3 is put which is similar to Γ_2 (and hence similar to Γ_1) but $1/\sqrt{2}$ times smaller Γ_2 (i.e., half the size of Γ_1). Proceeding in this manner we obtain a sequence of similar curves converging to a point at the distance $d\sqrt{2}/(\sqrt{2}-1)$ from Γ_1. The kth curve Γ_k of our sequence is capable of bounding two disc-type minimal surfaces, one of area $2^{1-k}A$, the other of area $2^{1-k}B$. Considering all the possibilities, the assembly of curves $\Gamma_1, \Gamma_2, \ldots$ will bound uncountably many—disconnected—minimal surfaces, each characterized by the value $\xi = (\varepsilon_1, \varepsilon_2, \ldots)$ of a dyadic number ($\varepsilon_k = 0$ or 1) and having total area

$$\sum_{k=1}^{\infty} 2^{1-k}\left[A + \varepsilon_k(B-A)\right]$$

$$= 2A + 2(B-A)\sum_{k=1}^{\infty} \varepsilon_k 2^{-k} = 2A + 2(B-A)\xi.$$

The areas of these disconnected minimal surfaces therefore can attain every value in the interval $[2A, 2B]$. Our disconnected "contour" bounding infinitely many disconnected minimal surfaces is now changed into a rectifiable Jordan curve by connecting bridges whose size decreases in the same geometric progression as those of the curves $\Gamma_1, \Gamma_2, \ldots$. Figure 5 serves as an illustration. If it is supposed that the validity of the bridge theorem remains unimpaired even for the case of infinitely many contours strung together by infinitely many bridges, then an example of a rectifiable Jordan curve bounding uncountably many solutions of Plateau's problem

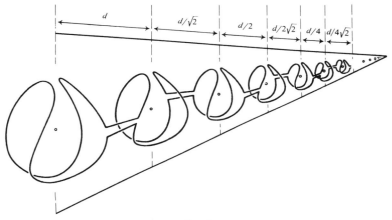

$$\text{FIG. 5}$$

has been found. Moreover, the areas of these surfaces presumably could assume every value in a certain interval.

The heuristic considerations outlined here have been a great challenge, and attempts at rigorous proofs abound. In a specific version (Γ_j of class $C^{2,\alpha}$, S_j strictly area minimizing without branch points on Γ_j) the bridge theorem has been established by W. H. Meeks and S. T. Yau (1982).

It must be pointed out that the preceding remarks concern the solutions of Plateau's problem, that is, 2-dimensional disc-type minimal surfaces in 3-space. The situation is different for minimal surfaces with several boundaries, for nonorientable minimal surfaces, for minimal surfaces in higher dimensional spaces and for higher dimensional minimal varieties.

In the foregoing sections we have commented on a number of questions relating to the classical Plateau problem. The developments of the last decades have of course gone far beyond this problem. Douglas envisaged a general setting which he formulated in 1939 as follows:

Given any riemannian manifold R in the most general sense of the term, i.e., any two-dimensional topological variety for which there is defined in the neighborhood of each point a local conformal representation on a circle. R may then have any finite or infinite number of boundaries, and any topological structure whatever—i.e., any finite or infinite type of connectivity. It may also have either character of orientability, i.e., one- or two-sidedness. Given also any point-set Γ in n-dimensional euclidean space which is a topological image of the total boundary C of R. Γ may consist of any finite or infinite number of Jordan curves, together with their limit points; or it may be some more general type of point set. A definite sense of description is associated with each component of Γ, carried over from C, which may be supposed oriented so that R is on the left. To determine the existence of a minimal surface M topologically equivalent to R and bounded by Γ.

The study of the Riemann surfaces which are stipulated here as parameter manifolds and the variation of their conformal parameters will no doubt gain considerable clarification through the results of Teichmüller theory. One of Douglas's early discoveries was the theorem that two linked Jordan curves are always capable of bounding a ring-type minimal surface; see Figure 6.

Free boundary problems have been considered by Courant and his school. Here the solution surfaces are bounded, wholly or in part, by fixed supporting surfaces. The regularity behavior of the free boundary portions, or trace, has posed intriguing questions. For an analytic support, the trace of an area minimizing surface is analytic as well (H. Lewy, 1951). Again, there is also a Kellogg-type theorem (K. H. Goldhorn, S. Hildebrandt, W. Jäger, J. C. C. Nitsche, J. E. Taylor). The fact that this theorem is true not only for area minimizing solutions, but also for stationary surfaces, has been proved in 1981 by M. Grüter, S. Hildebrandt, J. C. C. Nitsche and by G. Dziuk. A situation where the free boundary contains a cusp is depicted in Figure 7. Experimental evidence leads to still other boundary value problems. All can be summarized in the following way: "To determine and to study surfaces of least or stationary area and of a prescribed or unspecified topological type, which are bounded by a configuration consisting of fixed contours, supporting surfaces and moveable threads of given length." It is interesting to note that a surface of least area always pulls a moveable part of its boundary into the shape of a space curve of constant curvature.

Much attention has recently been given to the so-called obstacle problems, generalizations to higher dimensions of old calculus of variations problems: To determine a surface S of least area with prescribed boundary which lies entirely on one side of a fixed point set. The questions of interest concern existence, regularity, behavior

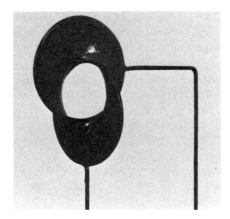

FIG. 6*

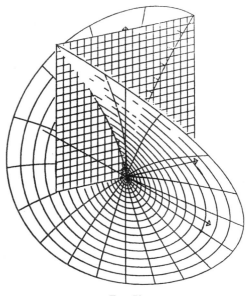

FIG. 7*

*Figure 6 gives another view of the model depicted in [25], p. 531. Figure 7 was kindly prepared by Dr. I. Haubitz at the Rechenzentrum of the University Würzburg.

of $\mathbb{S}$ near the obstacle as well as the nature of the coincidence set in the points of which $\mathbb{S}$ is attached to the obstacle. In a simple example, the surface $\mathbb{S}$ is assumed to allow a nonparametric representation $z = z(x, y)$ over a domain Ω in the (x, y)-plane satisfying $z|_{\partial\Omega} = 0$. The obstacle condition is expressed by the inequality $z(x, y) \geqslant \psi(x, y)$, where $\psi(x, y)$ is negative on $\partial\Omega$, but positive somewhere in Ω. Thin obstacles are those for which $\psi > 0$ only over a one-dimensional set (by the general maximum principle proved in Part 2, Section 4, to be nontrivial, this set must have positive linear Hausdorff measure). If $\partial\Omega$ is analytic and strictly convex, and if $\psi(x, y)$ is an analytic convex (as viewed from above) function, then the coincidence set is bounded by a regular analytic Jordan curve (D. Kinderlehrer, 1973).

In certain situations, the classical approaches to the boundary value problems mentioned are fraught with shortcomings. The solutions obtained may not be embeddings; they may lead to surfaces having transversal self-intersections, or possibly branch points—phenomena which are not observed in soap film experiments. One of the causes for this is inherent in the concept of parameter surfaces which proves ill-suited for many purposes, as it prejudges the topological type of the ultimate solution surface. For example, a knotted Jordan curve in 3-space cannot bound an embedded disc. An actual soap film through Γ will take another shape, often that of a Moebius band. The attack of Plateau's problem and various of its extensions in a more general framework was made possible by dealing with these problems from a set theoretic point of view. Initial success came from the work of A. S. Besicovitch, E. R. Reifenberg, H. Federer, et al., in the decade 1950–60. The theory of normal and integral currents was developed by Federer and W. Fleming in 1960. A setting often found more suitable for the description of physical experiments is provided by Almgren's theory of varifolds. This theory has also been applied to the mathematical treatment of aggregates of minimal surfaces—stable systems of minimal surfaces spanned into a frame. Detailed studies of such systems were already carried out by E. Lamarle, a collaborator of Plateau, in the 1860's. Two instructive examples, aggregates in a tetrahedral and a cubic frame, are depicted in Figures 8, 9, and 10.

FIG. 8

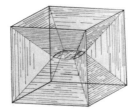

FIG. 9

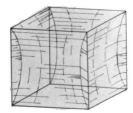

FIG. 10

Let us assume that a solution of a boundary value problem has been found in the form of a current or a varifold. One should then set one's sight on obtaining a full description of this structure, primarily on the demonstration that it is really as close to being a classical surface as possible. It turns out that two kinds of points on the solution structure must be distinguished, regular points and singular points. In the neighborhood of a regular point the current is manifold-like: That part of the current which is contained in a sufficiently small ball about one of its regular points is in fact a surface portion in the classical sense. This regular picture is disturbed near the singular points, and any information concerning the "singular set" becomes of interest. For the simplest case of Plateau's problem (for 2-dimensional surfaces in 3-space), the singular set is empty, and the solution structure is therefore everywhere regular. Globally it may be quite wild, possessing a number of, or even infinitely many handles. Locally, however, it is a smooth regular manifold. In more general cases, particularly if we are considering Plateau's problem in its higher-dimensional versions, the situation is more complex. It is known that the dimension of the singular set is lower than that of the solution structure. Precisely how voluminous it can be under the most unfavorable circumstances is a difficult question. For absolutely area minimizing (not for unstable) structures of codimension one, i.e., n-dimensional currents in $\mathbb{R}^{n+1}$, a remarkably satisfactory answer has been provided by the work of Fleming, DeGiorgi, Almgren, Federer, et al. Here the singular set will be empty as long as $n \leqslant 7$. For $n > 7$, the Hausdorff dimension of the singular set cannot be larger than $n - 7$. This interior regularity result has been complemented in 1979 by R. Hardt and L. Simon with a boundary regularity result: If the boundary of an absolutely area minimizing current $\mathcal{S}$ is locally a connected, oriented, embedded $C^{1,\alpha}$ manifold ($0 < \alpha < 1$), then $\mathcal{S}$ is locally a connected, embedded $C^{1,\alpha}$-hypersurface with boundary. ($\mathcal{S}$ has no boundary singularities, even though $\mathcal{S}$ may have an interior singular set of Hausdorff dimension $n - 7$.)

Figures 8, 9, 10 illustrate the two kinds of singularities which can appear in the stable aggregates of minimal surfaces:

(1) branch lines, or liquid edges, along which exactly three sheets meet under mutually equal angles of $120°$;
(2) vertices in which six surfaces and four liquid edges come together. Any pair of these edges includes the same angle $\alpha = 109.47°$ ($\cos \alpha = -1/3$).

All other hypothetically conceivable singularities are impossible. These facts which had been, somewhat heuristically, ascertained by Lamarle have been established by J. E. Taylor in 1976. Taylor also proved the interior local $C^{1,\alpha}$-regularity of area minimizing configurations. Higher interior regularity of the liquid edges was subsequently proved by Nitsche (C^∞, 1977) and Kinderlehrer-Nirenberg-Spruck (C^ω, 1978).

Many of the recent advances concerning minimal surfaces were unrelated to Plateau's problem. We mention here two. The first concerns a concrete isoperimetric inequality for multiply connected minimal surfaces. Assume that $\mathcal{S}$ is a ring-type minimal surface in $\mathbb{R}^3$ bounded by two disjoint Jordan curves Γ_1 and Γ_2. Note that $\mathcal{S}$ is not required to minimize area and that no assumptions are made about the regularity behavior, beyond continuity, of the surface $\mathcal{S}$ near its boundary. There exists a conformal mapping leading to a representation of $\mathcal{S}$ in the form $\mathcal{S} = \{\mathfrak{x} = \mathfrak{x}(u, v); (u, v) \in \overline{R}\}$ over the closure of a ring domain

$$R = \left\{ u, v; 0 < r_1^2 < u^2 + v^2 < r_2^2 < \infty \right\},$$

where the position vector is of class $C^2(R) \cap C^o(\overline{R})$ and satisfies in R the conditions $\Delta \mathfrak{x} = 0$, $\mathfrak{x}_u^2 = \mathfrak{x}_v^2$, $\mathfrak{x}_u \mathfrak{x}_v = 0$. Moreover, the boundary components $u^2 + v^2 = r_j^2$ ($j = 1, 2$) of R are mapped by $\mathfrak{x}(u, v)$ onto the curves Γ_j in a topological manner. Nitsche proved in 1965 that the area $A = A[\mathcal{S}]$ is finite for rectifiable curves Γ_j, and that $6A \leqslant (L_1 + L_2)^2$. Here L_1 and L_2 denote the lengths of Γ_1 and Γ_2, respectively. The sharp inequality $4\pi A \leqslant (L_1 + L_2)^2$ was established by Osserman and Schiffer in 1975. In fact, these authors proved that

$$(L_1 + L_2)^2 - 4\pi A > 2L_1 L_2(1 - \log 2)$$

and, if the curves Γ_1 and Γ_2 cannot be separated by a hyperplane, $L_1^2 + L_2^2 \geqslant 4\pi A$, i.e.,

$$(L_1 + L_2)^2 - 4\pi A \geqslant 2L_1 L_2.$$

It would be interesting to determine the largest constant κ for which the inequality

$$(L_1 + L_2)^2 - 4\pi A \geqslant \kappa L_1 L_2$$

is valid in the general case. (For a catenoid bounded by two circles in parallel planes, $\kappa = 1.6006$.)

Another study deals with surfaces of class $\mathcal{S}(h)$, that is, doubly-connected minimal surfaces $\mathcal{S}$ in the slab $-h < z < h$ of (x, y, z)-space which are intersected by each plane $z = c$, $-h < c < h$, in a starshaped curve $\mathcal{S}_c$. Denote by k_{min} and k_{max} the minimum and maximum curvature of the curve $\mathcal{S}_o$ (intersection of $\mathcal{S}$ with the plane $z = 0$) and by L its length. There is a number $q_o > 0$ and a function $m(q)$ defined for $q \geqslant q_0$ and satisfying the conditions $0 < m(q) < 1$, $\lim_{q \to \infty} m(q) = 1$, such that, for $h \geqslant q_0$,

$$\frac{k_{min}}{k_{max}} \geqslant m\left(\frac{h}{L}\right).$$

This inequality shows quantitatively that the "longer" the surface $\mathcal{S}$, the closer it will be to a catenoid. Surfaces of class $\mathcal{S}(\infty)$ must be catenoids (Nitsche, 1963).

Mention should also be made of another approach to the theory of "meromorphic minimal surfaces"—mappings from the complex plane into $\mathbb{R}^3$ by a vector with components which are the real parts of meromorphic functions satisfying relation (1.10)—in the papers of E. F. Beckenbach and his school. The aim here is an adaptation and generalization of Nevanlinna theory.

The preceding paragraphs contain repeated references to higher-dimensional minimal manifolds in higher-dimensional spaces. It appears requisite to conclude this exposition with a brief discussion of such manifolds. A slight change in the notation is in order.

Let $x = (x_i)$, $i = 1, 2, \ldots, n$, and $u = (u_\alpha)$, $\alpha = 1, 2, \ldots, m > n$, denote points in $\mathbb{R}^n$ and $\mathbb{R}^m$, respectively. Let $u(x): \Omega \to \mathbb{R}^m$ be a (sufficiently often) differentiable map from a domain $\Omega \subset \mathbb{R}^n$ onto a set $\mathcal{S}$ in $\mathbb{R}^m$. It is assumed that this map is bijective and regular, so that the Jacobian $(\partial u_\alpha / \partial x_i)$ has maximal rank n or, what is the same, that $g = \det(g_{ij}) > 0$. Here the $g_{ij}(x)$ are the coefficients of the first fundamental form,

$$g_{ij}(x) = \sum_{\alpha=1}^{m} \frac{\partial u_\alpha}{\partial x_i} \frac{\partial u_\alpha}{\partial x_j}.$$

The elements of the inverse matrix $(g_{ij})^{-1}$ will be denoted by g^{ij}. The bijectivity assumption may have to be dropped in general situations; it ensures that the submanifold $\mathbb{S}$ is embedded in the ambient space $\mathbb{R}^m$.

At every point p of $\mathbb{S}$, the n vectors $\partial u/\partial x_i$ span the n-dimensional tangent space $T_p(\mathbb{S})$ to $\mathbb{S}$ at p. The orthogonal complement of $T_p(\mathbb{S})$ in $\mathbb{R}^m$ is the $(m-n)$-dimensional normal space $N_p(\mathbb{S})$ to $\mathbb{S}$ at p. Every vector F in $\mathbb{R}^m$ can be decomposed into its tangential component $F^T \in T_p(\mathbb{S})$ and its normal component $F^N \in N_p(\mathbb{S})$.

The integral

$$V[\mathbb{S}] = \int_\Omega \sqrt{g(x)}\, dx_1 \dots dx_n$$

expresses the n-dimensional volume of $\mathbb{S}$. In order to investigate the change of this volume associated with a variation of the mapping $u(x)$, we consider a one-parameter family of mappings $u(x; t)$: $\Omega \times I \to \mathbb{R}^m$, where I is some interval on the real line containing the origin, such that $u(x; 0) = u(x)$. For each value of the parameter t we define the quantities $g_{ij}(x; t), g(x; t)$ as before. The volume of the image $\mathbb{S}(t)$ of Ω under the mapping $u(x; t)$ is

$$V[\mathbb{S}(t)] = \int_\Omega \sqrt{g(x; t)}\, dx_1 \dots dx_n,$$

and the expression $\dfrac{d}{dt} V[\mathbb{S}(t)]\Big|_{t=0}$ denotes the first variation of this volume. Stationary maps $u(x)$ are those for which the first variation vanishes, regardless of the special one-parameter family $u(x; t)$ in which they are embedded. A computation shows that the stationary character of a map is expressed by the condition that the mean curvature vector H of $\mathbb{S}$,

$$H = \left(\sum_{i,j=1}^n g^{ij} \frac{\partial^2 u}{\partial x_i\, \partial x_j} \right)^N,$$

vanish in all points of $\mathbb{S}$. (As in the simpler cases discussed earlier, the normal vector field H measures how the embedded manifold $\mathcal{S}$ is curved relative to the ambient space $\mathbb{R}^m$.) $\mathcal{S}$ is called *minimal* if $H \equiv 0$.

For any regular bijective mapping $u(x)$: $\Omega \to \mathcal{S} \subset \mathbb{R}^m$, the Laplace-Beltrami operator of $\mathcal{S}$, applied to any C^2-function ϕ: $\mathcal{S} \to \mathbb{R}$ (there is of course a weak version also) is defined by

$$\Delta\phi = \frac{1}{\sqrt{g}} \sum_{i,j=1}^{n} \frac{\partial}{\partial x_i}\left(\sqrt{g}\, g^{ij} \frac{\partial\phi}{\partial x_j}\right)$$

$$= \sum_{i,j=1}^{n} g^{ij}\left\{\frac{\partial^2\phi}{\partial x_i \partial x_j} - \sum_{k=1}^{n} \Gamma_{ij}^k \frac{\partial\phi}{\partial x_k}\right\}, \qquad (1.16)$$

where the Christoffel symbols Γ_{ij}^k are equal to

$$\Gamma_{ij}^k = \sum_{l=1}^{n} g^{kl}\Gamma_{ij,l}, \quad \Gamma_{ij,l} = \frac{1}{2}\left(\frac{\partial g_{jl}}{\partial x_i} + \frac{\partial g_{il}}{\partial x_j} - \frac{\partial g_{ij}}{\partial x_l}\right). \quad (1.17)$$

From the relations

$$\frac{\partial g_{jl}}{\partial x_i} = \sum_{\alpha=1}^{m}\left\{\frac{\partial^2 u_\alpha}{\partial x_i \partial x_j}\frac{\partial u_\alpha}{\partial x_l} + \frac{\partial^2 u_\alpha}{\partial x_i \partial x_l}\frac{\partial u_\alpha}{\partial x_j}\right\}$$

and similar expressions for $\partial g_{il}/\partial x_j$, $\partial g_{ij}/\partial x_l$, it is seen that

$$\sum_{\alpha=1}^{m} \frac{\partial^2 u_\alpha}{\partial x_i \partial x_j}\frac{\partial u_\alpha}{\partial x_l} = \Gamma_{ij,l}.$$

Also,

$$\sum_{\alpha=1}^{m}\sum_{k=1}^{n} \Gamma_{ij}^k \frac{\partial u_\alpha}{\partial x_k}\frac{\partial u_\alpha}{\partial x_l} = \sum_{k=1}^{n} g_{lk}\Gamma_{ij}^k = \Gamma_{ij,k}.$$

Thus,

$$\sum_{\alpha=1}^{m} \frac{\partial u_\alpha}{\partial x_l}\Delta u_\alpha = 0$$

for $l = 1, 2, \ldots, n$. It follows that the vector Δu is contained in the normal space $N_p(\mathcal{S})$ at each point $p = u(x)$ of $\mathcal{S}$ and, moreover, that

$$\Delta u = (\Delta u)^N = H. \tag{1.18}$$

(1.18) implies that $\mathcal{S}$ is minimal if and only if each coordinate function $u_\alpha(x)$ is harmonic on $\mathcal{S}$.

For a suitable choice of coordinates in $\mathbb{R}^m$, the minimal submanifold $\mathcal{S}$ can be represented locally in the "nonparametric" form

$$\begin{aligned} u_i &= x_i & i &= 1, 2, \ldots, n \\ u_\alpha &= u_\alpha(x_1, \ldots, x_n) & \alpha &= n+1, \ldots, m. \end{aligned} \tag{1.19}$$

The quantities g_{ij}, g^{ij}, g are computed with respect to the parameters $x_1, \ldots, x_n$. Since $\partial^2 u_k / \partial x_i \partial x_j = 0$ for $k = 1, 2, \ldots, n$, the conditions for $\mathcal{S}$ to be an n-dimensional minimal submanifold of $\mathbb{R}^m$ are expressed by the $m - n$ equations

$$\sum_{i,j=1}^n g^{ij} \frac{\partial^2 u_\alpha}{\partial x_i \partial x_j} = 0, \quad \alpha = n+1, \ldots, m. \tag{1.20}$$

In this way the local study of minimal submanifolds in Euclidean space $\mathbb{R}^m$ is equivalent to the study of the elliptic system of partial differential equations (1.20).

Particular attention has been given to the case of minimal hypersurfaces (of codimension one) in a nonparametric representation $u = u(x_1, \ldots, x_n)$ in $(x_1, \ldots, x_n, u)$-space. Here the condition of minimality is reduced to the single elliptic differential equation

$$\sum_{i,j=1}^n g^{ij} \frac{\partial^2 u}{\partial x_i \partial x_j} = 0, \tag{1.21}$$

where

$$g_{ij} = \delta_{ij} + \frac{\partial u}{\partial x_i} \frac{\partial u}{\partial x_j}, \quad g^{ij} = \delta_{ij} - \frac{1}{g} \frac{\partial u}{\partial x_i} \frac{\partial u}{\partial x_j},$$

$$g = 1 + \sum_{i=1}^n \left(\frac{\partial u}{\partial x_i} \right)^2 \equiv W^2. \tag{1.22}$$

(1.21) is equivalent to

$$\sum_{i=1}^{n} \frac{\partial^2 u}{\partial x_i^2} - \frac{1}{W^2} \sum_{i,j=1}^{n} \frac{\partial u}{\partial x_i} \frac{\partial u}{\partial x_j} \frac{\partial^2 u}{\partial x_i \partial x_j} = 0 \qquad (1.23)$$

or

$$\sum_{i=1}^{n} \frac{\partial}{\partial x_i} \left(\frac{u_i}{W} \right) = 0. \qquad (1.24)$$

Here we have used the abbreviation $u_i = \partial u / \partial x_i$; we also write $u_{ij} = \partial^2 u / \partial x_i \partial x_j$.

A minimal hypersurface $u = u(x_1, \ldots, x_n)$ supports functions with useful properties. A calculation shows, for instance, that

$$\Delta \frac{1}{W} = -W^{-2} \left\{ \Delta W - \frac{2}{W} \nabla(W, W) \right\}$$

$$= -W^{-7} \left\{ W^4 \sum_{i,j} u_{ij}^2 - 2W^2 \sum_i \left(\sum_j u_j u_{ij} \right)^2 + \left(\sum_{i,j} u_i u_j u_{ij} \right)^2 \right\}$$

$$= -W^{-1} \left\{ \nabla \left(\frac{1}{W}, \frac{1}{W} \right) + \sum_i \nabla \left(\frac{u_i}{W}, \frac{u_i}{W} \right) \right\}$$

$$= -W^{-1} \sum_i \kappa_i^2.$$

The symbol ∇ in this formula denotes Beltrami's first differential operator,

$$\nabla(\phi, \psi) = \sum_{i,j} g^{ij} \frac{\partial \phi}{\partial x_i} \frac{\partial \psi}{\partial x_j},$$

and $\kappa_1, \kappa_2, \ldots, \kappa_n$ are the principal curvatures of our hypersurface, that is, the roots of the characteristic equation $\det |\kappa W g_{ij} - u_{ij}| = 0$. The expression in the curled bracket above is seen to be positive unless $u_{ij} \equiv 0$. For the function $h(x) = \log W$ we have $-W\Delta(1/W) = \Delta h - \nabla(h, h)$, so that

$$h = \log W > 0, \quad \Delta h \geq \nabla(h, h), \qquad (1.25)$$

with equality holding if and only if $u_{ij} \equiv 0$. Inequality (1.25) will be used in Section 2.2.

If the minimal hypersurface $\mathbb{S}$ is defined by a relation $\Phi(x_\alpha) = 0$, grad $\Phi \neq 0$, where we set $x_{n+1} = u$, then the condition of minimality becomes

$$\sum_{\alpha=1}^{n+1} \frac{\partial}{\partial x_\alpha} \left(\frac{\Phi_\alpha}{\sqrt{\sum_\beta \Phi_\beta^2}} \right) = 0 \; (\mathrm{mod}\, \Phi) \qquad (1.26)$$

or

$$\left(\sum_\alpha \Phi_{\alpha\alpha} \right)\left(\sum_\beta \Phi_\beta^2 \right) - \sum_{\alpha,\beta} \Phi_{\alpha\beta}\Phi_\alpha\Phi_\beta = 0 \; (\mathrm{mod}\, \Phi). \qquad (1.27)$$

Among the special minimal hypersurfaces are the minimal cones, which may have a singularity at their vertex. A cone in $\mathbb{R}^{n+1}$ is a union of rays emanating from the origin; it is uniquely determined by its intersection with the sphere $S^n: x_1^2 + x_2^2 + \cdots + x_{n+1}^2 = 1$. If $\Phi(x_\alpha)$ is a (irreducible) homogeneous polynomial satisfying (1.27), then the equation $\Phi(x_\alpha) = 0$ defines an algebraic minimal cone.

The study of minimal cones has been inspired by the measure-theoretic approach to minimal varieties. For the regularity theory it is important to find conditions under which a minimal cone has no singularity at the origin and is, in fact, a linear space. One may also ask how complicated a minimal cone can be. The determination of all quadratic minimal cones is easily carried out. Since equation (1.27) is invariant under orthogonal transformations, the defining function for such a cone may be assumed from the outset to have the form $\Phi = \sum_\alpha a_\alpha x_\alpha^2$. If one substitutes this function into (1.27) (subject to the condition $\Phi = 0$) and relabels the coordinates, if necessary, one finds the n possibilities

$$(n-p)\left(x_1^2 + \cdots + x_p^2 \right) - (p-1)\left(x_{p+1}^2 + \cdots + x_{n+1}^2 \right) = 0,$$

$$1 \leqslant p \leqslant n. \qquad (1.28)$$

The cone with equation (1.28) is invariant with respect to the action of the group $SO(p) \times SO(n+1-p)$. The cases $p=1$ and $p=n$

lead to hypersurfaces. The only minimal cones in $\mathbb{R}^3$ are planes through the origin. In $\mathbb{R}^4$ one has the minimal cone

$$x_1^2 + x_2^2 - x_3^2 - x_4^2 = 0, \tag{1.29}$$

but there are many others.

Algebraic cones of higher degree are harder to come by. A number of such cones have been determined by W. Y. Hsiang using Lie group methods. Whether there are transcendental minimal cones, as well as other questions seem to be open. One is reminded of the investigations a hundred years ago of real two-dimensional (nonplane) algebraic minimal surfaces by S. Lie, L. Henneberg, et al., which led to results such as the following:

(1)   The smallest class for such a minimal surface is 5. (There are algebraic minimal surfaces of class 3 and 4, but they are not real.)
(2)   The sum of degree and class is not smaller than 14.
(3)   The degree of all algebraic minimal surfaces of class 5 is 15.
(4)   If there are algebraic minimal surfaces of a degree smaller than 9, then their degree must be 6 and their class must be 9.

The foregoing developments could have been generalized by replacing the Euclidean space $\mathbb{R}^m$, or both spaces $\mathbb{R}^n$ and $\mathbb{R}^m$, with Riemannian manifolds. This would lead to a setting for the discussion of minimal submanifolds of spheres and other special Riemannian spaces. There is an extensive literature on such manifolds, including the work of F. J. Almgren, E. Calabi, B. Y. Chen, E. DeGiorgi, W. Y. Hsiang, H. B. Lawson, T. Otsuki, J. Simons, T. Takahashi, et al. Particularly interesting results have been obtained for the case that the ambient space is the m-dimensional sphere S^m (endowed with the standard metric). In $S^3: x_1^2 + x_2^2 + x_3^2 + x_4^2 = 1$, there is the Clifford torus of genus one,

$$\left\{ x_1 = \frac{1}{\sqrt{2}} \cos t_1, x_2 = \frac{1}{\sqrt{2}} \sin t_1, x_3 = \frac{1}{\sqrt{2}} \cos t_2, \right.$$

$$\left. x_4 = \frac{1}{\sqrt{2}} \sin t_2; 0 \leqslant t_1, t_2 \leqslant 2\pi \right\}. \tag{1.30}$$

In 1968, Lawson proved that there exist compact orientable 2-dimensional minimal surfaces without boundary, of arbitrary genus, embedded in S^3.

In Euclidean space, there cannot exist compact minimal submanifolds without boundary. Here the minimal submanifolds of geometric interest are the compact ones with boundary and the open (noncompact) ones which are complete in the induced metric.

General questions concerning compact minimal submanifolds of a Riemannian space have recently attracted much attention. Minimal immersions of 2-spheres and other 2-dimensional manifolds have been studied by E. Calabi, H. B. Lawson, T. Otsuki, C. K. Peng, J. T. Pitts, J. Sacks, R. Schoen, T. Takahashi, C. L. Terng, K. Uhlenbeck, S. T. Yau and others. The general program consists of two parts. First, the existence of an object which is minimal in a suitably general sense—a stationary integral varifold—is established. The proof that this object is actually an embedded smooth (or smooth, with the exception of a singular set) minimal submanifold, depends on the results of a far-reaching regularity theory. In 1979–80 Pitts announced the following remarkable results:

> Let M be an $(n + 1)$-dimensional, smooth, compact Riemannian manifold without boundary. There exists a nonempty, compact, embedded, n-dimensional minimal submanifold $\mathbb{S} \subset M$ without boundary. $\mathbb{S}$ is smooth except, possibly, in the points of an exceptional set. The ν-dimensional Hausdorff measure of this singular set vanishes for $\nu > n - 7$. (For $n \leqslant 6$, $\mathbb{S}$ is smooth without exception.)

This is a vast generalization of the well-known theorem of Ljusternik and Fet (1951) that an arbitrary compact manifold always supports a closed geodesic.

The well-defined intersection properties of two-dimensional minimal surfaces discussed earlier have made the latter a useful tool in the newest studies of low-dimensional topology, especially of 3-manifolds, by M. Freedman, J. Hass, W. Jaco, W. H. Meeks, R. Schoen, P. Scott, L. Simon, S. T. Yau.

There is an important relation between minimal cones in $\mathbb{R}^{n+1}$ and minimal submanifolds of the unit sphere $S^n \subset \mathbb{R}^{n+1}$: A cone $\Phi(x_1, \ldots, x_{n+1}) = 0$, with vertex at the origin, is a minimal hypersurface in $\mathbb{R}^{n+1}$ (regular, except at the vertex) if, and only if, its intersection with the unit sphere $S^n: x_1^2 + \cdots + x_{n+1}^2 = 1$ is a minimal hypersurface in S^n. For example, the minimal cone (1.29) intersects the sphere S^3 in the Clifford torus (1.30).

It should be noted that the term minimal cone does not imply that a cone is actually area-minimizing. There may exist variations of the part of the cone in the interior of S^n, keeping the boundary (intersection of the cone with S^n) fixed, which decrease the n-dimensional volume. In fact, such variations can always be constructed for $n \leqslant 6$. In 1968 J. Simons showed that the situation changes when the dimension reaches the value $n = 7$. The 7-dimensional cone

$$x_1^2 + x_2^2 + x_3^2 + x_4^2 = x_5^2 + x_6^2 + x_7^2 + x_8^2 \qquad (1.31)$$

in $\mathbb{R}^8$ (case $n = 7$, $p = 4$ of (1.28)), for instance, is stable, that is, small deformations, which keep the boundary in S^7 fixed, increase volume. Bombieri, Giusti, and DeGiorgi proved subsequently that the cone (1.31) realizes the absolute minimum of volume for the given boundary. Thus, here is an example of a solution for Plateau's problem in $\mathbb{R}^8$ which is not everywhere regular. As for the size of the set of possible singularities, H. Federer showed in 1970 that the Hausdorff dimension of the singular set for an n-dimensional area minimizing hypersurface in $\mathbb{R}^{n+1}$ cannot exceed the value $n - 7$. (Area minimizing hypersurfaces of dimension $n \leqslant 6$ are analytic manifolds.)

The question about singularities of minimal varieties becomes complex in the case of higher codimension. The following is an example, due to Lawson and Osserman, of a 4-dimensional, nonparametric minimal cone in $\mathbb{R}^7$ (set $u_5 = f_1, u_6 = f_2, u_7 = f_3$ in (1.19)):

$$f_1(x_1, x_2, x_3, x_4) = \frac{\sqrt{5}}{2}|x|^{-1}(x_1^2 + x_2^2 - x_3^2 - x_4^2)$$

$$f_2(x_1, x_2, x_3, x_4) = \sqrt{5}|x|^{-1}(x_1 x_3 + x_2 x_4) \qquad (1.32)$$

$$f_3(x_1, x_2, x_3, x_4) = \sqrt{5}|x|^{-1}(x_2 x_3 - x_1 x_4).$$

Here $|x| = (x_1^2 + x_2^2 + x_3^2 + x_4^2)^{1/2}$. The Lawson-Osserman cone is contained in the 6-dimensional cone

$$5(x_1^2 + x_2^2 + x_3^2 + x_4^2) - 4(x_5^2 + x_6^2 + x_7^2) = 0$$

in $\mathbb{R}^7$; the latter is not minimal. While the authors' discovery of the minimal cone (1.32) was a result of their theoretical arguments, the reader might attempt to verify by direct computation that the functions f_1, f_2, f_3 in (1.32) satisfy the system (1.20). (Such a verification has been accomplished with the help of the MACSYMA computer program.) (1.32) is also the example of a global Lipschitz solution to (1.20) which is not smooth.

The expression (1.28) for the quadratic minimal cones of codimension one suggests that one might also seek minimal hypersurfaces $\Phi(x_1,\ldots,x_{n+1}) = 0$ in $\mathbb{R}^{n+1}$ which are invariant under the group $SO(p) \times SO(n+1-p)$. For such surfaces, $\Phi = \phi(\xi, \eta)$, where

$$\xi = \left(x_1^2 + \cdots + x_p^2\right)^{1/2}, \quad \eta = \left(x_{p+1}^2 + \cdots + x_{n+1}^2\right)^{1/2}, \quad 1 \leqslant p \leqslant n.$$

The differential equation (1.27) becomes

$$\phi_\eta^2 \phi_{\xi\xi} - 2\phi_\xi \phi_\eta \phi_{\xi\eta} + \phi_\xi^2 \phi_{\eta\eta} + \left(\frac{p-1}{\xi}\phi_\xi + \frac{n-p}{\eta}\phi_\eta\right)(\phi_\xi^2 + \phi_\eta^2) = 0$$

$$(\mathrm{mod}\,\phi). \quad (1.33)$$

In order to rid ourselves of the qualifier $(\mathrm{mod}\,\phi)$, we assume the relation $\phi(\xi, \eta) = 0$ to be parametrized in the form $\xi = \xi(t), \eta = \eta(t)$. This leads to the condition

$$\xi'\eta'' - \xi''\eta' + (\xi'^2 + \eta'^2)\left[(p-1)\frac{\eta'}{\xi} - (n-p)\frac{\xi'}{\eta}\right] = 0. \quad (1.34)$$

It is an interesting observation that (1.34) is identical with the Euler-Lagrange equation for the parametric variational problem

$$\delta \iint E(\xi, \eta)\sqrt{\xi'^2 + \eta'^2}\, dt = 0, \quad E(\xi, \eta) = \xi^{p-1}\eta^{n-p}. \quad (1.35)$$

In other words: The "curve" $\xi = \xi(t)$, $\eta = \eta(t)$ describing our minimal hypersurface is a geodesic for the metric

$$ds^2 = \xi^{p-1}\eta^{n-p}\left(d\xi^2 + d\eta^2\right). \qquad (1.36)$$

As an example, let us look for minimal hypersurfaces in $\mathbb{R}^4$ which have the Clifford torus (1.30) as boundary. Then we have to find all geodesics for the metric

$$ds^2 = \xi\eta\left(d\xi^2 + d\eta^2\right)$$

in the first quadrant of the (ξ, η)-plane which start at the point $(1/\sqrt{2}, 1/\sqrt{2})$ and end at $\xi\eta = 0$. There is exactly one such geodesic ending at $\xi = \eta = 0$, namely $\xi = \eta$. This geodesic corresponds to the cone (1.29). A discussion by Bombieri shows that there are infinitely many other admissible geodesics: The Clifford torus bounds infinitely many 3-dimensional minimal hypersurfaces.

There are also nonparametric minimal hypersurfaces $u = u(x_1, \ldots, x_n)$ in $\mathbb{R}^{n+1}$ which can be represented in the form $u = f(\xi, \eta)$ where

$$\xi = \left(x_1^2 + \cdots + x_p^2\right)^{1/2}, \quad \eta\left(x_{p+1}^2 + \cdots + x_n^2\right)^{1/2}, \quad 1 \leqslant p \leqslant n-1.$$

For such surfaces, the differential equation (1.23) becomes

$$\left(1 + f_\eta^2\right)f_{\xi\xi} - 2f_\xi f_\eta f_{\xi\eta} + \left(1 + f_\xi^2\right)f_{\eta\eta}$$

$$+ \left(\frac{p-1}{\xi}f_\xi + \frac{n-p-1}{\eta}f_\eta\right)\left(1 + f_\xi^2 + f_\eta^2\right) = 0. \qquad (1.37)$$

PART 2

1. Boundary behavior of minimal surfaces. We shall present here a short proof for the theorem regarding the boundary regularity of solutions of Plateau's problem which was formulated in Part 1 (pp. 84–85). The proof will be given for 2-dimensional surfaces in

$\mathbb{R}^3$; only superficial changes are necessary for such surfaces in $\mathbb{R}^{n+1}$, $n > 2$. It applies to all solution surfaces and is not restricted to the area-minimizing or stable ones.

Let $\mathfrak{S} = \{\mathfrak{x} = \mathfrak{x}(u, v); (u, v) \in \bar{P}\}$ be a solution of Plateau's problem, as defined in Part 1 (p. 83). The position vector

$$\mathfrak{x}(u, v) = \{x(u, v), y(u, v), z(u, v)\}$$

can be represented with the help of three analytic functions in the form

$$\mathfrak{x}(u, v) = \text{Re}\{g_1(w), g_2(w), g_3(x)\}.$$

The functions $g_j(w)$ satisfy Monge's relation (1.10). A motion of the (x, y, z)-coordinate system results in a linear transformation of the vector $\{g_1(w), g_2(w), g_3(w)\}$ into a new vector $\{\tilde{g}_1(w), \tilde{g}_2(w), \tilde{g}_3(w)\}$ by an orthogonal matrix with real-valued elements,

$$\tilde{g}_j(w) = a_j + \sum_{k=1}^{3} a_{jk} g_k(w).$$

Thus, if the mth derivatives ($m \geqslant 1$) of the functions $g_j(w)$ satisfy an inequality, say $|g_j^{(m)}(w)| \leqslant M$, for certain values of w, then the corresponding derivatives of the functions $\tilde{g}_j(w)$ satisfy a similar inequality, namely $|\tilde{g}_j^{(m)}(w)| \leqslant \sqrt{3}\, M$, for the same values of w. The same relation holds for the bounds on the differences $|g_j(w) - g_j(w_0)|$. We shall make repeated use of this simple remark, referring to it later as "argument $\mathcal{Q}$". Setting $u + iv = w = \rho e^{i\theta}$, we shall interchangeably use the notations $\mathfrak{x}(u, v)$, or $\mathfrak{x}(w)$, or $\mathfrak{x}(\rho e^{i\theta})$—whichever is most convenient.

The following simple facts have been known for some time:

(a) If Γ is rectifiable, then the position vectors of all solutions of Plateau's problem for Γ are uniformly equicontinuous.

(b) If Γ is rectifiable, then the derivatives $g_j'(w)$ for $j = 1, 2, 3$ belong to the Hardy class H_1, which implies, among other things, the absolute continuity of the vector $\mathfrak{x}(e^{i\theta})$ so that

$\mathfrak{x}_\theta(e^{i\theta})$ exists almost everywhere. Furthermore, $|\mathfrak{x}_\theta(e^{i\theta})| > 0$ for almost all values of θ.

(c) If Γ is a regular curve of class C^1, then the vector $\mathfrak{x}(w)$ belongs to class $C^{0,\beta}(\bar{P})$ for every exponent $\beta < 1/2$.

In particular,

$$|\mathfrak{x}(e^{i\theta_2}) - \mathfrak{x}(e^{i\theta_1})| \leqslant \mathcal{C}|\theta_2 - \theta_1|, \quad 0 \leqslant \beta < 1/2. \quad (2.1.1)$$

Here the constant $\mathcal{C}$ depends on β and on the properties of Γ, but it is independent of the individual solution of Plateau's problem. We shall encounter more such constants later on and shall denote them successively by $\mathcal{C}_1, \mathcal{C}_2, \ldots$.

For a review of these facts, the reader may consult paragraphs 316, 297; 321, 322; 328, 329, respectively, of [25].

$\Gamma = \{\mathfrak{x} = \mathfrak{z}(\tau); 0 \leqslant \tau \leqslant 2\pi\}$ is said to be a regular Jordan curve of class $C^{m,\alpha}$ ($m \geqslant 1$, $0 < \alpha < 1$) if the (periodic) position vector $\mathfrak{z}(\tau)$ has m continuous derivatives and satisfies the inequalities $|\mathfrak{z}'(\tau)| > 0$ and

$$|\mathfrak{z}^{(m)}(\tau_2) - \mathfrak{z}^{(m)}(\tau_1)| \leqslant M|\tau_2 - \tau_1|^\alpha$$

for all τ, τ_1, τ_2. Such a curve has the following property: Consider an arbitrary point on Γ and choose a coordinate system in which this point is the origin while the tangent to Γ becomes the x-axis. There exist positive constants d and $\mathcal{C}_1$ such that the connected subarc of Γ containing the (new) origin and lying in the slab $|x| \leqslant d$ of (x, y, z)-space has a representation

$$y = \psi(x), \quad z = \chi(x), \quad |x| \leqslant d.$$

The functions $\psi(x)$ and $\chi(x)$ possess m continuous derivatives and satisfy the relations

$$\psi(0) = \psi'(0) = \chi(0) = \chi'(0) = 0$$

and

$$|\psi^{(k)}(x)| \leqslant \mathcal{C}_1, \quad |\chi^{(k)}(x)| \leqslant \mathcal{C}_1 \quad \text{for } k = 0, 1, \ldots, m$$

$$|\psi^{(m)}(x_2) - \psi^{(m)}(x_1)| \leqslant \mathcal{C}_1|x_2 - x_1|^\alpha,$$

$$|\chi^{(m)}(x_2) - \chi^{(m)}(x_1)| \leqslant \mathcal{C}_1|x_2 - x_1|^\alpha \quad (2.1.2)$$

for $-d \leqslant x, x_1, x_2 \leqslant d$. The constants d and $\mathcal{C}_1$ *are the same* for all points on Γ.

For the proof of the regularity theorem two lemmas are useful. In these lemmas a complex valued function $g(w)$ is considered which is analytic and bounded in $|w| < 1$, so that the radial limits $\lim_{\rho \to 1} g(\rho e^{i\theta}) = g(e^{i\theta})$ exist for almost all values of θ. We set $h(\theta) = \operatorname{Re} g(e^{i\theta})$. $h(\theta)$ is an integrable function.

LEMMA 1. *Assume that $h(\theta_0)$ is finite and that for almost all θ the inequalities $|h(\theta) - h(\theta_0) - \kappa \sin(\theta - \theta_0)| \leqslant M(\theta - \theta_0)^{\lambda}, \lambda > k - 1,$ in $|\theta - \theta_0| \leqslant \delta$ and $|h(\theta)| \leqslant m$ otherwise are valid, where $k \geqslant 1$ is an integer. Then, for $1/2 \leqslant \rho < 1$,*

$$
|g^{(k)}(e^{i\theta_0})| \leqslant \begin{cases} N(1-\rho)^{\lambda - k} & \text{if} \quad k - 1 < \lambda < k \\ N \log \dfrac{1}{1-\rho} & \text{if} \quad \lambda = k \\ N & \text{if} \quad \lambda > k. \end{cases}
$$

Here the constant N depends on κ, λ, δ, M, m only.

Proof. Since the real part of the function $g(w) + i\kappa e^{-i\theta_0} w$ is equal to $h(\theta) - \kappa \sin(\theta - \theta_0)$ for $w = e^{i\theta}$, it is sufficient to prove the lemma under the assumption $\kappa = 0$. Differentiating Schwarz's relation

$$
g(w) = i \operatorname{Im} g(0) + \frac{1}{2\pi} \int_{-\pi}^{+\pi} \frac{e^{i\phi} + w}{e^{i\phi} - w} h(\phi) \, d\phi,
$$

we obtain, for $w = \rho e^{i\theta}$,

$$
g^{(k)}(w) = \frac{k!}{\pi} \int_{-\pi}^{+\pi} \frac{e^{i\phi}}{(e^{i\phi} - w)^{k+1}} h(\phi) \, d\phi
$$

$$
= \frac{k!}{\pi} \int_{-\pi}^{+\pi} \frac{e^{i(\phi - k\theta)}}{(e^{i\phi} - \rho)^{k+1}} h(\phi + \theta) \, d\phi
$$

$$
= \frac{k!}{\pi} \int_{-\pi}^{+\pi} \frac{h(\phi + \theta) - h(\theta)}{(e^{i\phi} - \rho)^{k+1}} e^{i(\phi - k\theta)} \, d\phi.
$$

Thus

$$|g^{(k)}(\rho e^{i\theta_0})| \leqslant \frac{k!}{\pi} \int_{-\pi}^{+\pi} \frac{|h(\phi + \theta_0) - h(\theta_0)|}{(e^{i\phi} - \rho)^{k+1}} d\phi.$$

Since $\sin \alpha \geqslant 2\alpha/\pi$ for $0 \leqslant \alpha \leqslant \pi/2$, we have the estimate

$$|e^{i\phi} - \rho|^2 = 1 - 2\rho \cos \phi + \rho^2 = (1 - \rho)^2 + 4\rho \sin^2(\phi/2)$$

$$\geqslant (1 - \rho)^2 + 4\rho \phi^2/\pi^2 \geqslant (1 - \rho)^2 + 2\phi^2/\pi^2$$

if $\rho \geqslant 1/2$. Therefore, for $1/2 \leqslant \rho < 1$,

$$|g^{(k)}(\rho e^{i\theta_0})| \leqslant \frac{k!}{\pi} \left(\int_{-\pi}^{-\delta} + \int_{-\delta}^{+\delta} + \int_{+\delta}^{+\pi} \right) \frac{|h(\phi + \theta_0) - h(\theta_0)|}{\left[(1 - \rho)^2 + \frac{2\phi^2}{\pi^2} \right]^{k+1}} d\phi$$

$$\leqslant 4mk! \left(\frac{\pi}{\delta\sqrt{2}} \right)^{k+1} + Mk!\pi^\lambda 2^{(3-\lambda)/2}(1 - \rho)^{\lambda - k} \int_0^{\sqrt{2}\delta/\pi(1-\rho)} \frac{t^\lambda dt}{\sqrt{1 + t^2}^{k+1}}.$$

A crude estimation shows that

$$\int_0^{\sqrt{2}\delta/\pi(1-\rho)} \frac{t^\lambda dt}{\sqrt{1 + t^2}^{k+1}} \leqslant \begin{cases} \dfrac{C}{k - \lambda} & \text{if } k - 1 < \lambda < k, \\[2mm] C \log \dfrac{1}{1 - \rho} & \text{if } \lambda = k, \\[2mm] \dfrac{C}{\lambda - k}(1 - \rho)^{k - \lambda} & \text{if } \lambda > k. \end{cases}$$

The constant C depends on δ, λ, k only. The statement of the lemma follows from this.

LEMMA 2. *Assume that the derivative g'(w) satisfies for all $1/2 \leqslant \rho < 1$, $|\theta - \theta_0| \leqslant \delta < 1/2$ one of the following inequalities:*

$$|g'(\rho e^{i\theta})| \leqslant \begin{cases} N(1 - \rho)^{\lambda - 1} & \text{for } 0 < \lambda < 1, \\[2mm] N \log \dfrac{1}{1 - \rho}, \\[2mm] N. \end{cases}$$

Then, for all θ_1, θ_2 satisfying $|\theta_1 - \theta_0| \leqslant \delta$, $|\theta_2 - \theta_0| \leqslant \delta$,

$$|g(e^{i\theta_2}) - g(e^{i\theta_1})| \leqslant \begin{cases} C|\theta_2 - \theta_1|^\lambda, \\ C|\theta_2 - \theta_1| \log \dfrac{1}{|\theta_2 - \theta_1|}, \\ C|\theta_2 - \theta_1|, \end{cases}$$

respectively. The constant C depends on λ, δ, N only.

Proof. For θ_1, θ_2 in the interval $\theta_0 - \delta \leqslant \theta_1, \theta_2 \leqslant \theta_0 + \delta$ we have

$$g(e^{i\theta_2}) - g(e^{i\theta_1}) = \int_{\lambda_r} g'(w) \, dw.$$

Here λ_r, for a suitable $r < 1$, denotes the path consisting of the two radial segments with end points $re^{i\theta_1}$, $e^{i\theta_1}$ and $re^{i\theta_2}$, $e^{i\theta_2}$ and the circular arc connecting the points $re^{i\theta_1}$, $re^{i\theta_2}$. Therefore

$$|g(e^{i\theta_2}) - g(e^{i\theta_1})| \leqslant \int_r^1 |g'(\rho e^{i\theta_1})| d\rho + \left| r \int_{\theta_1}^{\theta_2} |g'(re^{i\phi})| d\phi \right|$$

$$+ \int_r^1 |g'(\rho e^{i\theta_2})| d\rho.$$

In the first case we find

$$|g(e^{i\theta_2}) - g(e^{i\theta_1})| \leqslant \frac{N}{\lambda}(1-r)^\lambda + |\theta_2 - \theta_1| r(1-r)^{\lambda-1} + \frac{N}{r}(1-r)^\lambda.$$

With the choice $r = 1 - \frac{1}{2}|\theta_2 - \theta_1| > 1/2$, this inequality becomes

$$|g(e^{i\theta_2}) - g(e^{i\theta_1})| \leqslant 2^{1-\lambda}\left(1 + \frac{N}{\lambda}\right)|\theta_2 - \theta_1|^\lambda.$$

The other assertions of Lemma 2 are derived in a similar manner. The lemma is proved.

Now let $\mathcal{S} = \{\mathfrak{x} = \mathfrak{x}(u, v); (u, v) \in \overline{P}\}$ be a solution of Plateau's problem for the regular Jordan curve Γ of class $C^{1,\alpha}$. Consider a point on ∂P and its image on Γ. Without impairing the generality of the developments to follow, we may assume that the points so chosen are $w = 1$ on ∂P and the origin of (x, y, z)-space on Γ, and that the x-axis coincides with the tangent to Γ in the latter point. According to (2.1.1), there is a positive angle $\theta_0 \leqslant \pi/2$ such that the arc $\{w = e^{i\theta}; |\theta| \leqslant \theta_0\}$ of ∂P is mapped into the slab $|x| \leqslant d$. Thus $y(e^{i\theta}) = \psi(x(e^{i\theta}))$ and $z(e^{i\theta}) = \chi(x(e^{i\theta}))$. The angle θ_0 does not depend on the location of the chosen point on ∂P. The derivative $x_\theta(e^{i\theta})$ is nonnegative (or nonpositive—without loss of generality we choose the first option) wherever it exists. Since $|\mathfrak{x}_\theta(e^{i\theta})| > 0$ almost everywhere on ∂P, by (b), the inequality $x_\theta(e^{i\theta}) > 0$ is satisfied for almost all values of θ in $|\theta| \leqslant \theta_0$.

It follows from (2.1.1), (2.1.2) that

$$|y(e^{i\theta}) - y(1)| = |\psi(x(e^{i\theta}))| \leqslant \mathcal{C}_1 |x(e^{i\theta})|^{1+\alpha} \leqslant \mathcal{C}_1 \mathcal{C}^{1+\alpha} |\theta|^{(1+\alpha)\beta}$$

$$(2.1.3)$$

for $|\theta| \leqslant \theta_0$ and $|y(e^{i\theta})| \leqslant L$ for all θ. Here L denotes the length of Γ. Similar estimates hold for the function $z(e^{i\theta})$.

It will be convenient to work with a specific value of β (smaller than $1/2$) for which none of the power products $(1 + \alpha)^n \beta$ ($n = 2, 3, \ldots$) equals 1. Denote by n_0 that integer for which $(1 + \alpha)^{n_0} \beta < 1$ and $(1 + \alpha)^{n_0 + 1} \beta > 1$. By Lemma 1, with $k = 1$, $\kappa = 0$, the derivatives $g_2'(w)$ and $g_3'(w)$ satisfy the inequalities

$$|g_2'(\rho)| \leqslant \mathcal{C}_2 (1 - \rho)^{(1+\alpha)\beta - 1}, \quad |g_3'(\rho)| \leqslant \mathcal{C}_2 (1 - \rho)^{(1+\alpha)\beta - 1}$$

$$\text{for } 1/2 \leqslant \rho < 1.$$

The constant $\mathcal{C}_2$ now depends exclusively on the properties of Γ. By Monge's relation (1.10), a similar inequality (with the constant $\sqrt{2}\,\mathcal{C}_1$) must hold for $g_1'(\rho)$ also. Remembering that $w = 1$ was no special point for our considerations, we can conclude by argument $\mathcal{Q}$ that the derivatives $g_j'(w)$ satisfy the inequality

$$|g_j'(w)| \leqslant \sqrt{6}\,\mathcal{C}_2 (1 - |w|)^{(1+\alpha)\beta - 1} \quad \text{for } 1/2 \leqslant |w| < 1.$$

From this it follows by Lemma 2 that

$$|g_j(e^{i\theta_2}) - g_j(e^{i\theta_1})| \leqslant \mathcal{C}_3 |\theta_2 - \theta_1|^{(1+\alpha)\beta}$$

for all θ_1, θ_2, so that the functions $g_j(w)$ are seen to belong to the regularity class $C^{0,(1+\alpha)\beta}(\bar{P})$.

Repeating our procedure we find successively that the functions $g_j(w)$ and the position vector $\mathfrak{x}(u, v)$ will satisfy in $\bar{P}$ Hölder conditions with exponents $(1 + \alpha)^2\beta, \ldots, (1 + \alpha)^{n_0}\beta$.

The next step, based on the estimate (2.1.3), leads to an application of Lemma 1 for the case $\lambda = (1 + \alpha)^{n_0+1}\beta$, $k = 2$, $\kappa = 0$. In view of (1.10) we find

$$|g_j''(\rho)| \leqslant \mathcal{C}_4 (1 - \rho)^{\lambda - 2} \quad \text{for} \quad j = 1, 2, 3 \quad \text{and} \quad 1/2 \leqslant \rho < 1.$$

(The estimate for $g''_1(\rho)$ is based on the preliminary observation that all three derivatives $g_j'(w)$ are bounded and requires special care; see [25], §344.)

Argument $\mathcal{Q}$ and Lemma 2 now assure us that the derivatives $g_j'(w)$ satisfy in $\bar{P}$ a Hölder condition with exponent $(1 + \alpha)^{n_0+1}\beta - 1$.

At this stage, we may differentiate the relation $y(e^{i\theta}) = \psi(x(e^{i\theta}))$ to obtain

$$y_\theta(e^{i\theta}) = \psi'(x(e^{i\theta})) x_\theta(e^{i\theta})$$

and

$$|y_\theta(e^{i\theta}) - y_\theta(1)| \leqslant \mathcal{C}_5 |x(e^{i\theta})|^\alpha \leqslant \mathcal{C}_6 |\theta|^\alpha.$$

Similarly, $|z_\theta(e^{i\theta}) - z_\theta(1)| \leqslant \mathcal{C}_6 |\theta|^\alpha$. As before, it can be concluded that the functions $g_j(w)$ as well as the vector $\mathfrak{x}(u, v)$ belong to the regularity class $C^{1,\alpha}(\bar{P})$. The proof of the regularity theorem for the case $m = 1$ is completed:

If Γ is a regular Jordan curve of class $C^{1,\alpha}(0 < \alpha < 1)$, then $\mathfrak{x}(u, v)$ belongs to the regularity class $C^{1,\alpha}(\bar{P})$. The norm $\|\mathfrak{x}\|_{1,\alpha}^{\bar{P}}$ can be estimated uniformly for all solutions of Plateau's problem.

Repeated application of our "bootstrap procedure" leads to the regularity theorem for the higher cases $m = 2, 3, \ldots$.

The proof of the boundary regularity theorem presented here is due to J. C. C. Nitsche.

2. Gradient estimates for nonparametric minimal hypersurfaces.

Let Ω be a domain in $\mathbb{R}^n$ and $u(x) = u(x_1, \ldots, x_n)$ a function in $C^2(\Omega)$ satisfying the minimal surface equation (1.21),

$$\mathfrak{L}[u] \equiv \sum_{i,j=1}^{n} g^{ij}\left(\frac{\partial u}{\partial x_k}\right)\frac{\partial^2 u}{\partial x_i \partial x_j} = 0.$$

This differential equation is elliptic; it follows from (1.22) that

$$\frac{1}{W^2}\sum_{i=1}^{n}\xi_i^2 \leqslant \sum_{i,j=1}^{n} g^{ij}\xi_i\xi_j \leqslant \sum_{i=1}^{n}\xi_i^2.$$

Thus, the operator $\mathfrak{L}[u]$ is uniformly elliptic in every subdomain of Ω in which the first derivatives of $u(x)$ are known to be bounded. In this case the basic results of DeGiorgi and Ladyzhenskaya–Ural'tseva lead to a Hölder condition for the first derivatives of u: Let Ω' be a domain compactly contained in Ω and denote by $d > 0$ the distance between Ω' and the boundary of Ω. There are positive constants $\alpha = \alpha(n, M)$, $\mathcal{C} = \mathcal{C}(n, M, d)$ depending on the arguments indicated (but *not* on the individual solution) such that

$$\|u\|_{1,\alpha}^{\Omega'} \leqslant \mathcal{C} \quad \text{if} \quad \|u\|_{1,0}^{\Omega} = \sup_{\Omega}|u| + \sup_{\Omega}\max_{k}\left|\frac{\partial u}{\partial x_k}\right| \leqslant M.$$

$$(2.2.1)$$

Once inequality (2.2.1) is established, the general theory of elliptic equations is applicable and regularity results as well as estimates for the higher derivatives can be obtained. It is an important fact, due to the structure of the minimal surface equation, that a bound for a solution $u(x)$ of the minimal surface equation in Ω implies a bound for the gradient of $u(x)$ in every compact subdomain of Ω. The proof of this gradient estimate is the goal of the present section.

For any point x_0 in Ω we have

$$W(x_0) \leqslant c_1 \exp\left\{ c_2 \frac{\sup\limits_{\Omega} u(x) - u(x_0)}{d} \right\}. \qquad (2.2.2)$$

Here $d = \text{dist}(x_0, \partial\Omega)$, and c_1, c_2 are universal constants depending only on n.

An immediate consequence of (2.2.2), obtained by applying (2.2.2) to the solution $v(x) = -u(x)$, is an estimate

$$W(x_0) \leqslant c_1 e^{c_2 u(x_0)/d} \qquad (2.2.3)$$

for positive solutions of (1.23).

As a special consequence of (2.2.2) we mention the following compactness theorem:

Let $\{u(x)\}$ be a set of C^2-solutions of the minimal surface equation in Ω, and assume that the $u(x)$ are uniformly bounded. There exists a subsequence $\{u^{(\nu)}(x)\}$ converging to a C^2 (in fact, an analytic) solution of the minimal surface equation in Ω. The convergence is uniform in every compact subdomain of Ω.

For the case $n = 2$, inequality (2.2.2) was first proved by R. Finn in 1954. In this case the precise value of the constants is $c_1 = 1$, $c_2 = \pi/2$. The proof for n-dimensional minimal graphs was accomplished in 1969 by Bombieri, DeGiorgi, and Miranda. The version of the proof presented in the following is due to Trudinger.

Without loss of generality it may be assumed that the origin $x = 0$ belongs to Ω and that $u(0) = 0$. We denote by $X = \{x_i, u\}$ the position vector in $\mathbb{R}^{n+1}$ of our surface $\mathcal{S} = \{u = u(x); \ x \in \Omega\}$ and by $|X| = (|x|^2 + u^2)^{1/2}$ its length. Here $|x|^2 = \Sigma x_i^2$. X is a function of $x_1, x_2, \ldots, x_n$. A simple computation shows that

$$\frac{\partial}{\partial x_j}|X|^p = p|X|^{p-2}(x_j + uu_j), \tag{2.2.4}$$

$$\sum_j g^{ij}\frac{\partial}{\partial x_j}|X|^p = p|X|^{p-2}\left\{ x_i + \frac{u_i}{W}(X\cdot\nu) \right\}, \tag{2.2.5}$$

$$\nabla(|X|, |X|) = 1 - \frac{(X\cdot\nu)^2}{|X|^2}, \tag{2.2.6}$$

$$\Delta|X|^{p+2} = (p+2)|X|^p\left\{ p + n - \frac{p}{|X|^2}(X\cdot\nu)^2 \right\}. \tag{2.2.7}$$

Here $u_i = \partial u/\partial x_i$ and $\nu = \{-u_i/W, 1/W\}$ stands for the normal vector of $\mathbb{S}$, so that $X\cdot\nu = (u - \Sigma x_i u_i)/W$. We also recall the relation

$$\sum_i \frac{\partial}{\partial x_i}(Wg^{ij}) = 0 \quad \text{for all } j = 1, \ldots, n \tag{2.2.8}$$

and note that

$$\phi\Delta\psi - \psi\Delta\phi = \frac{1}{W}\sum_{i,j}\frac{\partial}{\partial x_i}\left(Wg^{ij}[\phi\psi_j - \psi\phi_j]\right), \tag{2.2.9}$$

$$\Delta(\phi\psi) = \phi\Delta\psi + \psi\Delta\phi + 2\nabla(\phi, \psi), \quad \nabla(\phi, \psi) = \sum_{i,j}g^{ij}\phi_i\psi_j \tag{2.2.10}$$

for any pair of C^2-functions ϕ, ψ.

If $k(x) \in C_0^2(\Omega)$, i.e., $k(x) \in C^2(\Omega)$ has compact support in Ω, and $\phi(x) \in C^2(\Omega)$, it follows from (2.2.9) that

$$\int_\Omega \phi(\Delta k)W\,dx = \int_\Omega k(\Delta\phi)W\,dx. \tag{2.2.11}$$

We shall prove the estimate (2.2.2) only for the case $n > 2$ and select for use in (2.2.11) the test function

$$\phi(x) = \begin{cases} \dfrac{n}{2}\varepsilon^{2-n} + \left(1 - \dfrac{n}{2}\right)\varepsilon^{-n}|X|^2, & |X| \le \varepsilon, \\ |X|^{2-n}, & |X| \ge \varepsilon \end{cases},$$

where $\varepsilon > 0$. $\phi(x)$ is continuously differentiable in Ω, and

$$\Delta\phi = \begin{cases} n(2-n)\varepsilon^{-n}, & |X| \le \varepsilon \\ n(2-n)|X|^{-(n+2)}(X\cdot\nu)^2, & |X| \ge \varepsilon \end{cases}.$$

Obviously, the relation (2.2.11) is also valid with the function $\phi(x)$. For $k(x)$ we choose $k(x) = \xi(x)h(x)$. Here $h(x) = \log W(x) > 0$ and $\xi(x) \in C_0^2(\Omega)$ is a cut-off function satisfying $0 \le \xi \le 1$ in Ω and $\xi = 1$ near $x = 0$. With the notation Ω_ε for the set $\{x; x \in \Omega, |X| < \varepsilon\}$, (2.2.11) leads to

$$n(2-n)\varepsilon^{-n}\int_{\Omega_\varepsilon} kW\,dx + n(2-n)\int_{\Omega\setminus\Omega_\varepsilon} k|X|^{-(n+2)}(X\cdot\nu)^2 W\,dx$$

$$= \int_\Omega \phi(\Delta k)W\,dx$$

or

$$n(n-2)\varepsilon^{-n}\int_{\Omega_\varepsilon} kW\,dx = -\int_\Omega \phi(\Delta k)W\,dx$$

$$-\,n(n-2)\int_{\Omega\setminus\Omega_\varepsilon} k|X|^{-(n+2)}(X\cdot\nu)^2 W\,dx$$

$$\le -\int_\Omega \phi(\Delta k)W\,dx.$$

Since $k \geqslant 0$, and

$$\varepsilon^{-n} \int_{\Omega_\varepsilon} kW \, dx \leqslant \frac{1}{2n} \varepsilon^{-n} \int_{\Omega_\varepsilon} |X|^2 (\Delta k) W \, dx$$

$$- \frac{1}{2(n-2)} \varepsilon^{2-n} \int_{\Omega_\varepsilon} (\Delta k) W \, dx$$

$$- \frac{1}{n(n-2)} \int_{\Omega \setminus \Omega_\varepsilon} |X|^{2-n} (\Delta k) W \, dx,$$

we have $\lim_{\varepsilon \to 0} \int_{\Omega_\varepsilon} W \, dx = \omega_n$ (the volume of the unit sphere in $\mathbb{R}^{n+1}$) and thus find for $\varepsilon \to 0$,

$$h(0) \leqslant - \frac{1}{n(n-2)\omega_n} \int_\Omega \frac{\Delta k}{|X|^{n-2}} W \, dx. \qquad (2.2.12)$$

The integral on the right is a convergent improper integral.

It follows from (2.2.5), (2.2.8), (2.2.10) that

$$\Delta k + \Delta(\xi h) = \xi \Delta h + h \Delta \xi + 2\nabla(\xi, h)$$

$$W|X|^{2-n} \nabla(\xi, h) = \sum_{i,j} \frac{\partial}{\partial x_i} \left(\frac{hW g^{ij} \xi_j}{|X|^{n-2}} \right) - h \sum_{i,j} \frac{\partial}{\partial x_i} \left(\frac{W g^{ij} \xi_j}{|X|^{n-2}} \right)$$

and

$$\sum_{i,j} \frac{\partial}{\partial x_i} \left(\frac{W g^{ij} \xi_j}{|X|^{n-2}} \right) = \frac{W}{|X|^{n-2}} \Delta \xi + \sum_{i,j} W g^{ij} \xi_j \frac{\partial}{\partial x_i} (|X|^{2-n})$$

$$= \frac{W}{|X|^{n-2}} \Delta \xi - (n-2) \frac{W}{|X|^n} \sum_j \xi_j \left\{ x_j + \frac{u_j}{W} (X \cdot \nu) \right\},$$

so that

$$\int_\Omega \frac{\nabla(\xi, h)}{|X|^{n-2}} W \, dx = - \int_\Omega \frac{h \Delta \xi}{|X|^{n-2}} W \, dx$$

$$+ (n-2) \int \frac{h}{|X|^n} \left\{ \sum_j \xi_j \left[x_j + \frac{u_j}{W} (X \cdot \nu) \right] \right\} W \, dx$$

and finally

$$\int_{\Omega} \frac{\Delta k}{|X|^{n-2}} W \, dx = \int_{\Omega} \frac{\xi \Delta h}{|X|^{n-2}} W \, dx$$

$$- \int_{\Omega} h \left\{ \frac{\Delta \xi}{|X|^{n-2}} - \frac{2(n-2)}{|X|^n} \sum_j \xi_j \left[x_j + \frac{u_j}{W} (X \cdot \nu) \right] \right\} W \, dx.$$

Since $\Delta h \geq 0$, by (1.25), the inequality (2.2.12) implies

$$h(0) \leq \frac{1}{n(n-2)\omega_n} \int_{\Omega} h \left\{ \frac{\Delta \xi}{|X|^{n-2}} - \frac{2(n-2)}{|X|^n} \right.$$

$$\left. \times \sum_j \xi_j \left[x_j + \frac{u_j}{W} (X \cdot \nu) \right] \right\} W \, dx. \qquad (2.2.13)$$

We now fix $\xi(x)$ further as follows. Let $R > 0$ be a number such that $\Omega_R \subset \Omega$, and subject to additional conditions to be specified later. Choose a function $\phi(t) \in C^2[0, \infty)$ which satisfies the conditions

$$0 \leq \phi(t) \leq 1, \phi(t) = 1 \text{ for } 0 \leq t \leq R/2,$$

$$\phi(t) = 0 \text{ for } t \geq R, |\phi'(t)| \leq 4/R, |\phi''(t)| \leq 16/R^2.$$

Set $\xi(x) = \phi(|X|)$. From

$$\xi_i = \phi' \frac{\partial}{\partial x_i} |X|, \Delta \xi = \phi' \Delta |X| + \phi'' \nabla(|X|, |X|)$$

we find with the help of (2.2.4), (2.2.6), (2.2.7) that, for $R/2 \leq |X| \leq R$,

$$|\Delta \xi| \leq \frac{8(n+2)}{R^2}, \quad \left| \sum_j \xi_j \left[x_j + \frac{u_j}{W} (X \cdot \nu) \right] \right| \leq 4$$

and thus

$$\left| \frac{\Delta \xi}{|X|^{n-2}} - \frac{2(n-2)}{|X|^n} \sum_j \xi_j \left[x_j + \frac{u_j}{W} (X \cdot v) \right] \right| \leqslant \frac{2^{n+1}(5n-6)}{R^n} .$$

(2.2.13) then yields

$$h(0) \leqslant \frac{\mathcal{C}}{R^n} \int_{\Omega_R} hW \, dx. \qquad (2.2.14)$$

Here and later, $\mathcal{C}$ denotes some constant depending on n.

We denote by $S(R)$ the ball in

$$\mathbb{R}^n \colon \{x; \, |x| < R\},$$

by $C(R)$ the infinite cylinder in

$$\mathbb{R}^{n+1} \colon \{x, u; \, |x| < R\},$$

and by $C(R; M_1, M_2)$ the finite cylinder in

$$\mathbb{R}^{n+1} \colon \{x, u; \, |x| < R, \, M_1 < u < M_2\}.$$

Obviously, $\Omega_R \subset \Omega \cap S(R)$. Assume that R is so small that $S(3R) \subset \Omega$. (2.2.14) then implies

$$h(0) \leqslant \frac{\mathcal{C}}{R^n} \int_{C(R; -R, +R)} hW \, dx. \qquad (2.2.15)$$

In its weak formulation, the minimal surface equation (1.24) is equivalent to the condition that

$$\int_\Omega \frac{1}{W} \left(\sum_i u_i \phi_i \right) dx = 0 \qquad (2.2.16)$$

for all $\phi(x) \in C_0^1(\Omega)$. Let

$$[u]_{-R}^R = \begin{cases} R, & u \geqslant R \\ u, & -R \leqslant u \leqslant R \\ -R, & u \leqslant -R \end{cases}$$

and set $u^R = R + [u]_{-R}^R$. Let $\eta(x) \in C_0^1(S(2R))$ be a test function satisfying $0 \leqslant \eta(x) \leqslant 1$, $n(x) = 1$ for $x \in S(R)$, $(\Sigma \eta_i^2)^{1/2} \leqslant 2/R$. Now choose in (2.2.16) $\phi(x) = hu^R\eta$. We have

$$\sum_i u_i \phi_i = \sum_i h\eta u_i u_i^R + \sum_i u^R \eta u_i h_i + \sum_i u^R h u_i \eta_i$$

and

$$\int_\Omega \frac{1}{W}(\Sigma u_i \phi_i)\, dx = \int_{C(2R;\, -R, R)} \frac{\eta h}{W} \Sigma u_i^2\, dx$$

$$+ \int_{C(2R;\, -R, \infty)} \frac{u^R \eta}{W} \Sigma u_i h_i\, dx$$

$$+ \int_{C(2R;\, -R, \infty)} \frac{u^R h}{W} \Sigma u_i \eta_i\, dx.$$

Since $0 \leqslant u^R \leqslant 2R$, it follows that

$$\int_{C(R;\, -R, R)} \frac{h}{W} \Sigma u_i^2\, dx \leqslant 2R \int_{C(2R;\, -R, \infty)} \left(\eta\sqrt{\Sigma h_i^2} + h\sqrt{\Sigma \eta_i^2} \right) dx.$$

$$(2.2.17)$$

For an estimation of the integral $\int \eta\sqrt{\Sigma h_i^2}\, dx$, we shall use the inequality (1.25). Let $\psi(x) \in C_0^1(S(2R))$ be a nonnegative test function. Then

$$\int_{S(2R)} \psi^2 \nabla(h, h) W\, dx \leqslant \int_{S(2R)} \psi^2(\Delta h) W\, dx$$

$$= -2 \int_{S(2R)} \psi \nabla(h, \psi) W\, dx$$

$$\leqslant 2 \max_{S(2R)} \sqrt{\Sigma \psi_i^2} \int_{S(2R)} \psi\sqrt{\nabla(h, h)}\, W\, dx$$

and

$$\left(\int_{S(2R)} \psi\sqrt{\nabla(h, h)}\, W\, dx \right)^2 \leqslant \left(\int_{S(2R)} \psi^2 \nabla(h, h) W\, dx \right)\left(\int_{S(2R)} W\, dx \right).$$

Here we have used the inequalities

$$|\nabla(f,g)|^2 \leqslant \nabla(f,f)\nabla(g,g), \quad \frac{1}{W^2}\sum_i f_i^2 \leqslant \nabla(f,f) \leqslant \sum_i f_i^2.$$

$$(2.2.18)$$

Combining these inequalities, we obtain

$$\int_{S(2R)} \psi\sqrt{\sum h_i^2}\, dx \leqslant \int_{S(2R)} \psi\sqrt{\nabla(h,h)}\, W\, dx$$

$$\leqslant 2\max_{S(2R)} \sqrt{\sum\psi_i^2} \int_{S(2R)\cap\mathrm{supp}(\psi)} W\, dx.$$

We specify $\psi(x)$ further as follows. Let $\zeta(t)\in C_0^1(-2R, R+\max_{S(2R)}u(x))$ be a test function satisfying $\zeta(t)=1$ for $-R\leqslant t\leqslant \max_{S(2R)}u(x)$ and $0\leqslant\zeta(t)\leqslant 1$, $|\zeta'(t)|\leqslant 2/R$. Set $\psi(x)=\eta(x)\zeta(u(x))$ to obtain

$$\int_{S(2R)} \eta\sqrt{\sum_i h_i^2}\, dx \leqslant \int_{S(2R)} \psi\sqrt{\nabla(h,h)}\, W\, dx$$

$$\leqslant \frac{\mathcal{C}}{R}\int_{C(2R;\,-2R,\infty)} W\, dx. \qquad (2.2.19)$$

Since $\log W \leqslant W$, we also have

$$\int_{C(2R;\,-R,\infty)} h\sqrt{\sum\eta_i^2}\, dx \leqslant \frac{\mathcal{C}}{R}\int_{C(2R;\,-R,\infty)} W\, dx. \qquad (2.2.20)$$

Since $W^2=\sum u_i^2+1$, we conclude from (2.2.17), (2.2.19), (2.2.20) that

$$\int_{C(R;\,-R,R)} hW\, dx \leqslant \omega_n R^n + \int_{C(2R;\,-2R,\infty)} W\, dx. \qquad (2.2.21)$$

For an estimate of the integral on the right hand side in (2.2.21), we go back to (2.2.16), using now the test function

$$\phi(x) = \xi(x) \max[0, 2R + u(x)]$$

where $\xi(x) \in \overset{\circ}{C}_0^1(S(3R))$ is a function satisfying $\xi(x) = 1$ in $S(2R)$, $0 \leqslant \xi(x) \leqslant 1$ and $|\xi_i(x)| \leqslant 2/R$. We have $\phi_i = 0$ for $u(x) \leqslant -2R$ and otherwise

$$\frac{1}{W}\sum_i u_i \phi_i = \xi\left(W - \frac{1}{W}\right) + \left(2R + \max_{S(3R)} u\right)\sum_i \xi_i \frac{u_i}{W}.$$

Thus

$$\int_{C(2R; -2R, \infty)} W\, dx \leqslant \int_{C(3R; -2R, \infty)} \xi W\, dx = \int_{C(3R; -2R, \infty)} \frac{\xi}{W}\, dx$$

$$+ \left(2R + \max_{S(3R)} u\right)\int_{C(3R; -2R, \infty)} \Sigma \xi_i \frac{u_i}{W}\, dx$$

$$\leqslant R^n\left(\mathcal{C}_1 + \frac{\mathcal{C}_2}{R}\max_{S(3R)} u\right).$$

Combining this inequality with (2.2.15) and (2.2.21) and exponentiating, we find

$$W(0) \leqslant \mathcal{C}_1 \exp\left(\mathcal{C}_2 \max_{S(3R)} \frac{u}{R}\right) \leqslant \mathcal{C}_1 \exp\left(\mathcal{C}_2 \max_{\Omega} \frac{u}{R}\right).$$

The estimate (2.2.2) follows by a translation.

3. Bernstein's theorem. In 1916 S. Bernstein put forth the following theorem:

A C^2-solution $z(x, y)$ of the minimal surface equation (1.2) which is defined for all values of the variables x and y (we shall call $z(x, y)$ an entire solution) must be a linear function.

The surprising aspect of this theorem lies in the fact that the linearity of the solution $z(x, y)$ is a consequence of its mere existence over all of $\mathbb{R}^2$; no assumptions regarding growth or boundedness are required— a striking contrast to the situation for harmonic functions and for the solutions of more general linear elliptic differential equations. Over the years, numerous proofs for Bernstein's theorem have been discovered. We present here a particularly simple proof, due to J. C. C. Nitsche.

Let $z = z(x, y)$ be an entire solution of the minimal surface equation $\mathcal{L}[z] = 0$ and set as usual $p = z_x$, $q = z_y$, $W = \sqrt{1 + p^2 + q^2}$. Introduce the functions

$$\xi = \xi(x, y) = x + \int_{(0,0)}^{(x,y)} \frac{1}{W} \left[(1 + p^2) \, dx + pq \, dy \right] \equiv x + A(x, y),$$

$$\eta = \eta(x, y) = y + \int_{(0,0)}^{(x,y)} \frac{1}{W} \left[pq \, dx + (1 + q^2) \, dy \right] \equiv y + B(x, y).$$

$$(2.3.1)$$

The integrals in (2.3.1) are path independent, as is seen from the relations

$$\frac{\partial}{\partial y} \frac{1 + p^2}{W} - \frac{\partial}{\partial x} \frac{pq}{W} = -\frac{q}{W^3} \mathcal{L}[z] = 0,$$

$$\frac{\partial}{\partial y} \frac{pq}{W} - \frac{\partial}{\partial x} \frac{1 + q^2}{W} = \frac{p}{W^3} \mathcal{L}[z] = 0.$$

We claim that the mapping $(x, y) \to (\xi, \eta)$ increases distances. To show this, consider two different points (x_1, y_1) and (x_2, y_2) in the (x, y)-plane and apply the mean value theorem to the function

$$f(t) = h \left[A(x_1 + th, y_1 + tk) - A(x_1, y_1) \right]$$

$$+ k \left[B(x_1 + th, y_1 + tk) - B(x_1, y_1) \right],$$

where $h = x_2 - x_1$, $k = y_2 - y_1$. The result is $f(1) - f(0) = f'(\bar{t})$,

$0 < \bar{t} < 1$, or

$$(x_2 - x_1)[A(x_2, y_2) - A(x_1, y_1)]$$
$$+ (y_2 - y_1)[B(x_2, y_2) - B(x_1, y_1)]$$
$$= h^2 \bar{A}_x + hk(\bar{A}_y + \bar{B}_x) + k^2 \bar{B}_y$$
$$= \frac{1}{\overline{W}}\left[(1 + \bar{p}^2)h^2 + 2\bar{p}\bar{q}hk + (1 + \bar{q}^2)k^2\right]$$
$$\geq \frac{1}{\overline{W}}[h^2 + k^2] > 0.$$

Here $\bar{A}_x = A(\bar{x}, \bar{y})$, $\bar{p} = p(\bar{x}, \bar{y})$, etc., where

$$\bar{x} = x_1 + \bar{t}h, \bar{y} = y_1 + \bar{t}k.$$

Setting $A_i = A(x_i, y_i)$, $\xi_i = \xi(x_i, y_i)$, etc., we now see that

$$0 < (x_2 - x_1)(A_2 - A_1) + (y_2 - y_1)(B_2 - B_1) < (x_2 - x_1)$$
$$\times [(\xi_2 - \xi_1) - (x_2 - x_1)] + (y_2 - y_1)[(\eta_2 - \eta_1) - (y_2 - y_1)]$$

or

$$(x_2 - x_1)^2 + (y_2 - y_1)^2 < (x_2 - x_1)(\xi_2 - \xi_1) + (y_2 - y_1)(\eta_2 - \eta_1)$$
$$< \sqrt{(x_2 - x_1)^2 + (y_2 - y_1)^2}\sqrt{(\xi_2 - \xi_1)^2 + (\eta_2 - \eta_1)^2}$$
$$< (\xi_2 - \xi_1)^2 + (\eta_2 - \eta_1)^2.$$

In view of this dilating property, the mapping $(x, y) \to (\xi, \eta)$ is bijective, transforming the whole (x, y)-plane into the whole (ξ, η)-plane. The relation

$$ds^2 = dx^2 + dy^2 + dz^2 = (1 + p)^2 dx^2 + 2pq\, dx\, dy + (1 + q^2)\, dy^2$$
$$= \left(\frac{W}{W+1}\right)^2 (d\xi^2 + d\eta^2)$$

implies that ξ and η are isothermal parameters on our minimal surface $\mathbb{S}$: $z = z(x, y)$. The coordinate z is a harmonic function $\mathbb{S}$, satisfying the Laplace equation $z_{\xi\xi} + z_{\eta\eta} = 0$, and the complex valued combination $z_\xi - iz_\eta$ becomes thus an entire analytic function of the complex variable $\zeta = \xi + i\eta$. A simple computation shows that

$$\Omega(\zeta) = z_\xi - iz_\eta = \frac{p - iq}{1 + W}. \tag{2.3.2}$$

Obviously, $|\Omega(\zeta)| = (W - 1)/(W + 1) < 1$. By Liouville's theorem, $\Omega(\zeta)$ must be a constant; the same is true for $|\Omega(\zeta)|$ and W and ultimately for p and q. Bernstein's theorem is proved.

The function $\Omega(\zeta)$ in (2.3.2) has an obvious geometrical interpretation. The spherical image of $\mathbb{S}$ on the unit sphere, provided by the unit normal vector $\{X, Y, Z\} = (-p/W, -q/W, 1/W)$ can be transferred through the stereographic transformation $\sigma = X/(1 + Z)$, $\tau = Y/(1 + Z)$ from the north pole of the unit sphere to the equatorial (σ, τ)-plane. One finds that $\Omega = -(\sigma + i\tau)^{-1}$.

The proof of Bernstein's theorem can be broken into several parts. The minimal surface $\mathbb{S}$ is obviously complete in the differential geometric sense, i.e., each diverging curve in $\mathbb{S}$ has infinite length (H. Hopf-W. Rinow). As a consequence, it is seen that $\mathbb{S}$ is conformally of parabolic type (the image of the (x, y)-plane is the whole (ξ, η)-plane). The boundedness of the function $\Omega(\zeta)$ expresses the fact that the spherical image of $\mathbb{S}$ is contained in a hemisphere. The desired conclusion follows from an application of Liouville's theorem of complex function theory. The fact that $\mathbb{S}$ is given in a nonparametric representation might be considered of secondary importance. Rather, the main ingredients of Bernstein's theorem are:

   (i)   the completeness of $\mathbb{S}$,
   (ii) the conformally parabolic type of $\mathbb{S}$,
   (iii) the boundedness of the spherical image of $\mathbb{S}$.

Viewed in this way, Bernstein's theorem is an example for the interrelation between the global metric-geometric, the conformal and the topological properties of complete minimal surfaces.

In 1959, R. Osserman proved that if the spherical image of a complete minimal surface $\mathcal{S}$ omits a whole neighborhood of a point on the unit sphere, then $\mathcal{S}$ must be a plane. He subsequently refined the answer to the question how "thick" the complement on the unit sphere of the spherical image can be for a nonflat complete minimal surface by showing that this complement must be a point set of vanishing logarithmic capacity. In 1980 F. Xavier proved the surprising result that the complement can in fact contain at most ten points. For $k = 0, 1, 2, 3, 4$, there are explicit examples of non-plane complete minimal surfaces for which the spherical image omits precisely k given points on the unit sphere. Xavier has improved his theorem since, replacing the number ten by the number six (*Ann. of Math.*, **113** (1981), **115** (1982)). The cases $k = 5$ and $k = 6$ remain undecided today.

For a long time it remained a challenging question whether Bernstein's original theorem is true also for complete minimal graphs in $\mathbb{R}^{n+1}$, i.e., for n-dimensional entire minimal hypersurfaces in nonparametric representation $u = u(x_1, \ldots, x_n)$ in $\mathbb{R}^{n+1}$. Step by step, through the work of Fleming, DeGiorgi, Almgren, and Simons, an affirmative answer was given for all dimensions $n \leqslant 7$.

It should be noted again that the special attraction of Bernstein's theorem derives from the fact that no assumptions are made regarding the solution $u(x)$ beyond the assumption of its existence over all of $\mathbb{R}^n$. If further restrictions are put on $u(x)$, for instance the condition that grad u remain bounded for $|x| \to \infty$, then the minimal surface equation satisfied by $u(x)$ would become a uniformly elliptic equation, and the result would be of a different nature better described as a Liouville-type theorem. Under the assumption that $|\operatorname{grad} u(x)| \leqslant M < \infty$ in $\mathbb{R}^n$, J. Moser proved in 1961 that $u(x)$ must be a linear function. Later (after the discovery of the critical value $n = 8$ for the dimension) it was shown by Bombieri and Giusti that Moser's conclusion follows already from the assumption that all but at most seven partial derivatives of $u(x)$ remain bounded.

It was a surprise to the mathematical community when Bombieri, DeGiorgi, and Giusti discovered in 1965 that the theorem is false for $n \geqslant 8$. The three authors proved the existence of an entire solution $u(x_1, \ldots, x_8)$ of the minimal surface equation (1.24) which is not linear. In fact,

$$|u(x_i)| \geqslant |x_1^2 + x_2^2 + x_3^2 + x_4^2 - x_5^2 - x_6^2$$
$$- x_7^2 - x_8^2|[x_1^2 + \cdots + x_8^2]^{1/2},$$

so that

$$\limsup_{|x| \to \infty} \frac{|u(x_i)|}{|x|^3} = 1, \quad |x| = \sqrt{x_1^2 + \cdots + x_8^2}.$$

An explicit nonlinear entire solution $u(x)$ of (1.24) has not yet been found; nor is information available about the possible growth behavior for $|x| \to \infty$ of entire minimal graphs. So far, the function $u(x)$ is obtained as a limit of solutions which are "squeezed" in between a supersolution and a subsolution of the minimal surface equation (1.23).

In the following we shall present the essential steps of the construction although it will not be possible to reproduce some of the detailed computations. The solution will be sought in the form $u = f(\xi, \eta)$, where

$$\xi = \sqrt{x_1^2 + x_2^2 + x_3^2 + x_4^2}, \quad \eta = \sqrt{x_5^2 + x_6^2 + x_7^2 + x_8^2}.$$

The function $f(\xi, \eta)$ satisfies the differential equation (1.37) for $n = 8$, $p = 4$:

$$\mathcal{L}[f] \equiv \left(1 - f_\eta^2\right) f_{\xi\xi} - 2 f_\xi f_\eta f_{\xi\eta} + \left(1 + f_\xi^2\right) f_{\eta\eta}$$
$$+ 3\left(\frac{f_\xi}{\xi} + \frac{f_\eta}{\eta}\right)\left(1 + f_\xi^2 + f_\eta^2\right) = 0. \qquad (2.3.3)$$

With the use of "polar coordinates" $\xi = \rho \cos\theta$, $\eta = \rho \sin\theta$, (2.3.3) becomes

$$\mathcal{L}[f] \equiv \left(1 + \frac{1}{\rho^2} f_\theta^2\right) f_{\rho\rho} - \frac{2}{\rho^2} f_\rho f_\theta f_{\rho\theta} + \frac{1}{\rho^2}\left(1 + f_\rho^2\right) f_{\theta\theta}$$
$$+ \frac{1}{\rho} f_\rho\left(1 + f_\rho^2 + \frac{2}{\rho^2} f_\theta^2\right) + 6\left(\frac{f_\rho}{\rho} + \frac{1}{\rho^2} \cotan 2\theta f_\theta\right)$$
$$\times \left(1 + f_\rho^2 + \frac{1}{\rho^2} f_\theta^2\right) = 0. \qquad (2.3.4)$$

It seems difficult to find solutions of (2.3.3) or (2.3.4) in closed form. The occurrence of the factor $\cotan 2\theta$ in (2.2.4) suggests that we seek a solution of the structure $f = g(\rho)\cos 2\theta$. For such a function we find

$$\mathcal{L}\left[g(\rho)\cos 2\theta\right] = \cos 2\theta \left\{ \frac{1}{\rho^2}\left[-16g + 7\rho g' + \rho^2 g''\right]\right.$$

$$+ \frac{4g}{\rho^4}\left[-12g^2 + 8\rho gg' - 2\rho^2 g'^2 + \rho^2 gg''\right]$$

$$+ \frac{1}{\rho^4}\left[48g^3 - 32\rho g^2 g' - 8\rho^2 gg'^2\right.$$

$$\left.\left. + 7\rho^3 g'^3 - 4\rho^2 g^2 g''\right]\cos^2 2\theta\right\}.$$

In particular, if $g(\rho) = \rho^k$,

$$\mathcal{L}\left[\rho^k \cos 2\theta\right] = \{(k-2)(k+8)\rho^{k-2} - 4(k-3)(k-4)\rho^{3k-4}$$

$$+ (k-2)(k+2)(7k-12)\rho^{3k-4}\cos^2 2\theta\}\cos 2\theta.$$

It follows that the function

$$f^{(1)}(\xi, \eta) = \rho^3 \cos 2\theta = \sqrt{\xi^2 + \eta^2}\,(\xi^2 - \eta^2)$$

satisfies the inequalities

$$f^{(1)} > 0,\ \mathcal{L}\left[f^{(1)}\right] > 0 \text{ in the sector } Q_1 = \{\xi, \eta;\, 0 < \eta < \xi\}$$

$$f^{(1)} < 0,\ \mathcal{L}\left[f^{(1)}\right] < 0 \text{ in the sector } Q_2 = \{\xi, \eta;\, 0 < \xi < \eta\},$$

$$(2.3.5)$$

while $f^{(1)} = 0$ for $\xi = \eta$ and $f^{(1)}(\xi, \eta) = -f^{(1)}(\eta, \xi)$. The expressions

$$f_\xi^{(1)}/\xi = \left(3\xi^2 + \eta^2\right)/\sqrt{\xi^2 + \eta^2} \quad \text{and}$$

$$f_\eta^{(1)}/\eta = -\left(\xi^2 + 3\eta^2\right)/\sqrt{\xi^2 + \eta^2}$$

have continuous extensions to the closure $\overline{Q}$ of the quadrant $Q = \{\xi, \eta; \xi > 0, \eta > 0\}$. We set

$$F^{(1)}(x) = f^{(1)}(\xi, \eta) = |x|(x_1^2 + x_2^2 + x_3^2 + x_4^2 - x_5^2 - x_6^2 - x_7^2 - x_8^2).$$

The function $F^{(1)}(x)$ has continuous second derivatives in all of $\mathbb{R}^8$ and is real-analytic in $\mathbb{R}^8 \setminus \{0\}$. By construction,

$$\mathcal{L}[F^{(1)}] \geqslant 0 \text{ in } D_1 = \{x \in \mathbb{R}^8; x_5^2 + x_6^2 + x_7^2 + x_8^2 \leqslant x_1^2 + x_2^2 + x_3^2 + x_4^2\}$$

$$\mathcal{L}[F^{(1)}] \leqslant 0 \text{ in } D_2 = \{x \in \mathbb{R}^8; x_1^2 + x_2^2 + x_3^2 + x_4^2 \leqslant x_5^2 + x_6^2 + x_7^2 + x_8^2\}.$$

$$(2.3.6)$$

Moreover, $F^{(1)}(x)$ is invariant with respect to the group $SO(4) \times SO(4)$ which leaves the cone

$$C = \{x; x_1^2 + x_2^2 + x_3^2 + x_4^2 = x_5^2 + x_6^2 + x_7^2 + x_8^2\}$$

invariant.

Denote by $S_r = S_r(0)$ the ball $\{x; |x| < r\}$ in $\mathbb{R}^8$ of radius r with center at the origin. Let $U^{(r)}(x)$ be the solution of the minimal surface equation in S_r which has the values of $F^{(1)}(x)$ on ∂S_r. The existence and uniqueness of $U^{(r)}(x)$ is assured by the theory of minimal surfaces. It follows from the uniqueness that $U^{(r)}(x)$ is also invariant under the group $SO(4) \times SO(4)$ and that $U^{(r)}(x) = 0$ on the cone C. Thus it is seen that $U^{(r)}(x) = F^{(1)}(x)$ on $\partial(S_r \cap D_1)$ and on $\partial(S_r \cap D_2)$. Applying the maximum principle, we find

$$U^{(r)}(x) \geqslant F^{(1)}(x) \text{ for } x \in \overline{S}_r \cap \overline{D}_1$$

and

$$U^{(r)}(x) \leqslant F^{(1)}(x) \text{ for } x \in \overline{S}_r \cap \overline{D}_2,$$

i.e.,

$$|F^{(1)}(x)| \leqslant |U^{(r)}(x)| \quad \text{for } x \in \overline{S}_r. \qquad (2.3.7)$$

It is now necessary to have a second comparison function $F^{(2)}(x)$, very much like $F^{(1)}(x)$, with the help of which an inequality in the opposite direction can be ascertained, namely

$$|U^{(r)}(x)| \leqslant |F^{(2)}(x)| \quad \text{for } x \in \bar{S}_r. \tag{2.3.8}$$

This function will again be sought in the form $f^{(2)}(\xi, \eta)$, where $f^{(2)}(\xi, \eta)$ must have the following properties:

$$f^{(2)} > f^{(1)}, \, \mathcal{L}[f^{(2)}] < 0 \quad \text{in the sector } Q_1,$$

$$f^{(2)} < f^{(1)}, \, \mathcal{L}[f^{(2)}] > 0 \quad \text{in the sector } Q_2, \tag{2.3.9}$$

$$f^{(2)} = 0 \quad \text{for } \xi = \eta.$$

A function with these properties is much harder to come by than $f^{(1)}$; its determination is a tour de force which shall be omitted here. Suffice it to state that such a function $f^{(2)}(\xi, \eta)$, and the corresponding function $F^{(2)}(x, y)$, can be constructed.

Consider now the sequence of solutions of the minimal surface equation $U^{(r)}(x)$ for $r = 1, 2, \ldots$. Each $U^{(n)}(x)$ is defined in the sphere $S_n(0)$. It follows from (2.3.7), (2.3.8) that the $U^{(n)}(x)$ are uniformly bounded in $S_R(0)$ for all $n > R$:

$$|U^{(n)}(x)| \leqslant M = M(R) \quad \text{for } x \in S_R(0), \qquad n > R.$$

The essential point of this inequality is the independence of M from n. With the aid of the compactness theorem stated in Section 2.2 we see that a subsequence $\{U^{(n_v)}(x)\}$ can be extracted from the $U^{(n)}(x)$ which converges to a limit function $U(x)$ defined in all of $\mathbb{R}^8$. The convergence is uniform in every compact subdomain of $\mathbb{R}^8$, and $U(x)$ is a C^2-(in fact, an analytic) solution of the minimal surface equation in all of $\mathbb{R}^8$. Obviously,

$$|F^{(1)}(x)| \leqslant |U(x)| \leqslant |F^{(2)}(x)|$$

and in particular

$$\limsup_{|x| \to \infty} \frac{|U(x)|}{|x|^3} \geqslant \limsup_{|x| \to \infty} \frac{|F^{(1)}(x)|}{|x|^3} = 1.$$

Hence $U(x)$ cannot be a linear function. $U(x)$ is trivially also a solution of the minimal surface equation in $\mathbb{R}^k$ for $k \geqslant 8$.

It is therefore clear that Bernstein's theorem is false in $\mathbb{R}^n$ for $n \geqslant 8$.

The construction of the super and subsolutions $F^{(j)}(x)$ might appear somewhat haphazard; the specific choices lead to success, but are not absolutely necessary. One might be inclined to conclude that all entire solutions of the minimal surface equation must be of polynomial growth. This conjecture has not yet been confirmed.

Bernstein's original theorem illustrates the fact that the solutions of the minimal surface equation, a nonuniformly elliptic equation, behave strikingly different from harmonic functions and from the solutions of uniformly elliptic equations. Even within the realm of the minimal surface equation, another remarkable difference manifests itself at the dimension $n = 8$ (the critical value is $5 + 2\sqrt{2}$). It is interesting to note further structural changes as one moves from minimal graphs (of codimension one) to nonparametrically represented minimal submanifolds of higher codimension. These submanifolds appear as solutions of the elliptic system of partial differential equations (1.20). At the present time, our knowledge in this case is very limited although important insights have been obtained through studies of S. S. Chern and R. Osserman (1967); D. Fischer-Colbrie (1980), D. A. Hoffman and R. Osserman (1980); S. Hildebrandt, J. Jost, and K. O. Widman (1980) and others. A concrete special result is the following (R. Osserman): "Let $u_3(x_1, x_2)$ and $u_4(x_1, x_2)$ be entire solutions of the system (1.20). If the minimal surface $\mathcal{S}$ in $\mathbb{R}^4$ defined by these functions has finite total curvature, then $\mathcal{S}$ is either a plane, or $u_3 \pm iu_4$ is a complex polynomial in the variable $z = x_1 + ix_2$."

4. Removable singularities. The question of removable singularities provides another example for the salient differences between the minimal surface equation and uniformly elliptic differential equations. In 1951 L. Bers established the surprising fact that an isolated singularity for a solution $z(x, y)$ of the minimal surface equation (1.2) is ipso facto removable. Around the same time, R. Finn proved a similar theorem for a larger class of equations. It was later discovered that certain considerably larger exceptional sets were removable as well (DeGiorgi-Stampacchia, Nitsche, 1965):

THEOREM. *Let Ω be a domain in $\mathbb{R}^n$ and $E \subset \Omega$ a compact set of vanishing $(n-1)$-dimensional Hausdorff measure. Let $u(x) \in C^2(\Omega \setminus E)$ be a solution of the minimal surface equation (1.24) in $\Omega \setminus E$. It is possible to assign values for the function $u(x)$ in the points of E such that the extended function is a C^2-solution of the minimal surface equation in all of Ω.*

The vanishing of the $(n-1)$-dimensional Hausdorff measure $H_{n-1}(E)$ of E can be expressed in the following way: For each $\varepsilon > 0$ there is a finite covering of E by open spheres S_k of radii r_k $(k = 1, 2, \ldots, m(\varepsilon))$ such that $\sum r_k^{n-1} < \varepsilon$. With the aid of such a covering, we carry out a special construction. Set $\rho_\varepsilon = \min(r_k)$ and let S_k' and S_k'' be the open spheres, concentric with S_k, of radii $r_k' = r_k + \rho_\varepsilon$ and $r_k'' = r_k + 2\rho_\varepsilon$, respectively. The unions of the spheres S_k, S_k', S_k'' will be denoted by $D(\varepsilon), D'(\varepsilon), D''(\varepsilon)$, respectively.

Let $\phi(t) \in C^1$ be a nonnegative function with compact support in the interval $|t| \leqslant 1$ satisfying the condition $\int_{-\infty}^{\infty} \phi(t) \, dt = 1$, and set

$$\psi^{(\varepsilon)}(x) = \psi^{(\varepsilon)}(x_i) = \left(\frac{\sqrt{n}}{\rho_\varepsilon} \right)^n \int_{D'(\varepsilon)} \prod_{i=1}^{n} \phi\left(\frac{\sqrt{n}}{\rho_\varepsilon} (x_i - \xi_i) \right) d\xi.$$

The function $\psi^{(\varepsilon)}(x)$ is of class $C^1(\mathbb{R}^n)$, and we have

$$0 \leqslant \psi^{(\varepsilon)}(x) \leqslant 1, \, \psi^{(\varepsilon)}(x) = 1 \text{ on } D(\varepsilon), \, \psi^{(\varepsilon)}(x) = 0 \text{ exterior to } D''(\varepsilon),$$

and

$$|\psi_{x_j}^{(\varepsilon)}(x)| \leqslant 2M\sqrt{n}/\rho_\varepsilon,$$

where $M = \max |\phi'(\varepsilon)|$. The volume of the set $D''(\varepsilon) \setminus D(\varepsilon)$ is not larger than

$$\frac{\omega_n}{n} \sum_k \left[(r_k + 2\rho_\varepsilon)^n - r_k^n \right] \leqslant \frac{2\omega_n}{n} \rho_\varepsilon \sum_k \sum_{l=1}^{n} (r_k + 2\rho_\varepsilon)^{n-l} r_k^{l-1}$$

$$\leqslant \frac{2\omega_n}{n} \rho_\varepsilon \sum_k \sum_{l=1}^{n} 3^{n-l} r_k^{n-1}$$

$$\leqslant 3^n \frac{\omega_n}{n} \rho_\varepsilon \sum_k r_k^{n-1} \leqslant 3^n \frac{\omega_n}{n} \varepsilon \rho_\varepsilon.$$

We therefore find

$$\int_{\mathbb{R}^n} |\operatorname{grad} \psi^{(\varepsilon)}(x)| \, dx \leqslant \mathcal{C}\varepsilon, \quad \mathcal{C} = 2 \cdot 3^n M \omega_n.$$

The above construction produces a set of functions $\psi^{(\varepsilon)}(x) \in C_0^1(\mathbb{R}^n)$ satisfying the conditions $0 \leqslant \psi^{(\varepsilon)}(x) \leqslant 1$, $\psi^{(\varepsilon)}(x) = 1$ on E, and

$$\lim_{\varepsilon \to 0} \int_{\mathbb{R}^n} |\operatorname{grad} \psi^{(\varepsilon)}(x)| \, dx = 0. \tag{2.4.1}$$

We make another observation concerning solutions of the minimal surface equation. Let $u^{(1)}(x)$ and $u^{(2)}(x)$ be two such solutions and set as usual

$$p_i^{(l)} = u_{x_i}^{(l)}, \quad W^{(l)} = \left[1 + \sum_{i=1}^{n} \left(u_{x_i}^{(l)} \right)^2 \right]^{1/2}.$$

It is claimed that the expression

$$F(u^{(1)}, u^{(2)}) \equiv \sum_{i=1}^{n} \frac{\partial}{\partial x_i} \left[(u^{(2)} - u^{(1)}) \left(\frac{p_i^{(2)}}{W^{(2)}} - \frac{p_i^{(1)}}{W^{(1)}} \right) \right]$$

$$= \sum_{i=1}^{n} \left(p_i^{(2)} - p_i^{(1)} \right) \left(\frac{p_i^{(2)}}{W^{(2)}} - \frac{p_i^{(1)}}{W^{(1)}} \right) \tag{2.4.2}$$

satisfies the inequality

$$F(u^{(1)}, u^{(2)}) \geqslant \left(\max(W^{(1)}, W^{(2)}) \right)^{-3} \sum_{i=1}^{n} \left(p_i^{(2)} - p_i^{(1)} \right)^2 \geqslant 0. \tag{2.4.3}$$

In order to prove (2.4.3), set

$$p_i = p_i^{(1)} + t \left(p_i^{(2)} - p_i^{(1)} \right), \quad W = \left[1 + \sum p_i^2 \right]^{1/2}$$

and apply the mean value theorem, for fixed (x), to the function

$$f(t) = \sum_{i=1}^{n} \left(p_i^{(2)} - p_i^{(1)} \right)\left(\frac{p_i}{W} - \frac{p_i^{(1)}}{W^{(1)}} \right):$$

$$F(u^{(1)}, u^{(2)}) = f(1) - f(0) = f'(\tilde{t})$$

$$= \frac{1}{\tilde{W}} \sum_{i,j=1}^{n} \left(\delta_{ij} - \frac{\tilde{p}_i \tilde{p}_j}{\tilde{W}^2} \right)\left(p_i^{(2)} - p_i^{(1)} \right)\left(p_j^{(2)} - p_j^{(1)} \right)$$

$$\geqslant \frac{1}{\tilde{W}^3} \sum_{i=1}^{n} \left(p_i^{(2)} - p_i^{(1)} \right)^2.$$

Here

$$0 < \tilde{t} < 1, \tilde{p}_i = p_i^{(1)} + \tilde{t}\left(p_i^{(2)} - p_i^{(1)} \right), \tilde{W} = \left[1 + \Sigma \tilde{p}_i^2 \right]^{1/2}.$$

From the relation

$$\frac{d^2}{dt^2} W^2 = 2 \sum_{i=1}^{n} \left(p_i^{(2)} - p_i^{(1)} \right)^2 \geqslant 0,$$

it can be concluded that $\tilde{W} \leqslant \max(W^{(1)}, W^{(2)})$, and (2.4.3) is proved.

We shall prove the theorem on removable singularities for the case that Ω is the ball $B_R = \{x; |x| < R\}$. The proof is based on a remarkable general maximum principle:

Let $u^{(1)}(x), u^{(2)}(x) \in C^2(\Omega \setminus E)$ be two solutions of the minimal surface equation (1.24) in $\Omega \setminus E$ and assume that $\limsup[u^{(2)}(x) - u^{(1)}(x)] \leqslant M$ for approach of all boundary points of Ω. Then the inequality $u^{(2)}(x) - u^{(1)}(x) \leqslant M$ holds for all points of the set $\Omega \setminus E$.

For the proof of the maximum principle, it may be assumed that $M < \infty$. Choose two numbers M_1 and M_2 such that $M < M_1 < M_2 < \infty$. Choose $r < R$ so that $E \subset B_r$ and that $u^{(2)}(x) - u^{(1)}(x) < M_1$ for $x \in \partial B_r$. Let $\varepsilon_0 > 0$ be so small that the domains $D''(\varepsilon)$ are contained in B_r for $\varepsilon < \varepsilon_0$. Now consider the function

$$u(x) = \left[u^{(2)}(x) \right]_{u^{(1)}(x) + M_1}^{u^{(1)}(x) + M_2}$$

$$= \begin{cases} u^{(1)}(x) + M_2 & \text{if} \quad u^{(2)}(x) - u^{(1)}(x) > M_2 \\ u^{(2)}(x) & \text{if} \quad M_1 \leqslant u^{(2)}(x) - u^{(1)}(x) \leqslant M_2 \\ u^{(1)}(x) + M_1 & \text{if} \quad u^{(2)}(x) - u^{(1)}(x) < M_1. \end{cases}$$

For almost all points of $B_r \backslash D(\varepsilon)$, either $u_{x_i} = p_i = p_i^{(1)}$ or $u_{x_i} = p_i = p_i^{(2)}$. Thus, since

$$\eta^{(\varepsilon)}(x) = 1 - \psi^{(\varepsilon)}(x) = 1 \text{ on } B_r \backslash D''(\varepsilon),$$

we see from (2.4.3) that

$$I \equiv \int_{B_r \backslash D''(\varepsilon)} F(u^{(1)}, u) \, dx \leqslant \int_{B_r \backslash D(\varepsilon)} F(u^{(1)}, u^{(2)}) \eta^{(\varepsilon)} \, dx.$$

On the boundary of $B_r \backslash D(\varepsilon)$ we have either $\eta^{(\varepsilon)} = 0$ (if $x \in \partial D(\varepsilon)$) or $u - u^{(1)} - M_1 = 0$ (if $x \in \partial B_r$). An integration by parts using (2.4.2) then shows

$$I \leqslant - \int_{B_r \backslash D(\varepsilon)} \left(u - u^{(1)} - M_1 \right) \sum_{i=1}^{n} \left(\frac{p_i^{(2)}}{W^{(2)}} - \frac{p_i^{(1)}}{W^{(1)}} \right) \eta_{x_i}^{(\varepsilon)} \, dx$$

$$\leqslant 2(M_2 - M_1) \int_{B_r \backslash D(\varepsilon)} |\text{grad } \eta^{(\varepsilon)}| \, dx$$

$$\leqslant 2(M_2 - M_1) \int_{\mathbb{R}^n} |\text{grad } \psi^{(\varepsilon)}| \, dx.$$

Letting ε go to zero, we find that $I = 0$, and we conclude from (2.4.3) that $p_i = p_i^{(1)}$ almost everywhere in $B_r \backslash E$. Thus $u(x) - u^{(1)}(x) = \text{const.}$ in $B_r \backslash E$. The value of the constant must be M_1, because $u^{(2)}(x) - u^{(1)}(x) < M_1$ on ∂B_r, so that $u = u^{(1)} + M_1$ on ∂B_r. By definition, the relation $u(x) = u^{(1)}(x) + M_1$ is equivalent to the inequality $u^{(2)}(x) - u^{(1)}(x) < M_1$. Since M_1 was arbitrary, the maximum principle follows.

The maximum principle becomes false if the exceptional set E is too voluminous. This can be demonstrated by counterexamples. For an arbitrary compact set $E \subset \Omega$ of positive (or infinite) $(n-1)$-dimensional Hausdorff measure, which is contained in an $(n-1)$-dimensional hyperplane, J. C. C. Nitsche has constructed solutions $u^{(1)}, u^{(2)}$ which do not obey the maximum principle. It is an open question whether such solutions can be found for an arbitrary exceptional set.

The proof of the theorem on removable singularities is now easily completed if we assume, as we shall do here, that Dirichlet's problem for the minimal surface equation is uniquely solvable for spheres. Let B_r be chosen as before, and let $v(x)$ be the solution of the minimal surface equation in B_r such that $v = u$ on ∂B_r. By the general maximum principle applied to $\Omega = B_r$, we have $u(x) \equiv v(x)$ in $B_r \backslash E$. The solution $v(x)$ provides the desired, and unique extension of $u(x)$ to all of B_r. (Obviously, $u = v$ in $B_R \backslash B_r$.)

There is another version for the last step. Applying the general maximum principle to the difference $u(x + h) - u(x)$, where $h = (h_1, h_2, \ldots, h_n)$ is a small constant vector, it is seen that $u(x)$ belongs to the regularity class $C^{0,1}(\Omega \backslash E)$. Thus $u(x)$ is a weak solution of (1.24) in Ω, and therefore analytic in Ω.

An analysis shows that the tools of the proof are the following: gradient estimate, higher a priori estimates, maximum principle, solvability of Dirichlet's problem for spheres (or "H-convex" domains) with incomplete boundary data, regularity theory. These properties are shared by a broad class of equations so that generalizations of the removability theorem are possible, for instance, to quasi-linear elliptic equations $\operatorname{div} \mathcal{C}(\operatorname{grad} u) = \mathcal{B}(u)$ arising from a bounded conservation law ($|\mathcal{C}(\operatorname{grad} u)| \leq a < \infty$, J. Serrin, 1965).

5. Nonremovable singularities. As for the case of the general maximum principle, there are examples which show that the condition $H_{n-1}(E) = 0$ is optimal for the theorem on removable singularities, and one is led to the general *question*:

Given an arbitrary compact set $E = \mathbb{R}^n$ satisfying $H_{n-1}(E) > 0$, can one find an open neighborhood Ω of E and a solution $u(x) \in C^2(\Omega \backslash E)$ of (1.24) in $\Omega \backslash E$ for which E — or, at least, parts of E — are not removable?

as well as the further *problem*:

> *To study the behavior of $u(x)$ for approach of E.*

It should be noted that this question is of particular interest only for the case that E is "just barely too large" for the removability theorem to hold, the situation being obvious for too voluminous sets E —sets possessing interior points (solid obstacle problem) or $(n-1)$-dimensional hypersurfaces (thin obstacle problem).

We shall give here an affirmative answer to the question for the case of exceptional sets contained in hyperplanes. For this case the procedure is particularly transparent. The proof, with suitable modifications and additions, can be applied to exceptional sets contained in C^ω-hypersurfaces or in C^2-hypersurfaces for which the set of points of vanishing mean curvature has $(n-1)$-dimensional measure zero.

Assume now that the compact exceptional set E with $H_{n-1}(E) > 0$ is contained in a hyperplane, say the hyperplane $x_n = 0$, and that

$$E \subset S = \{x; \; |x| < R, \, x_n = 0\}.$$

Denote by B the half-ball

$$B = \{x; \; |x| < R, \, x_n > 0\}.$$

Consider coverings of E by open sets $0_1, 0_2, \ldots$ in S in such a way that $0_1 \supset \supset 0_2 \supset \supset \cdots \supset \supset E$, $H_{n-1}(\partial 0_j) = 0$ and $H_{n-1}(0_j \setminus E) \to 0$. Set $B_j = \partial B \setminus \bar{0}_j$.

Let $u_j(x)$ be the solution of the Dirichlet problem for (1.24) in B —call this problem ($*$)—assuming the boundary values

$$u_j(x) = 0 \text{ for } x \in B_j, \quad u_j(x) = 1 \text{ for } x \in 0_j.$$

Despite the incompleteness of the boundary data which are discontinuous and not prescribed on a set of vanishing $(n-1)$-dimensional Hausdorff measure, this Dirichlet problem has a unique solution in the regularity class $C^2(B) \cup C^0(B_j \cup 0_j)$. The sequence of the solutions $u_j(x)$ for $j = 1, 2, \ldots$ is monotonously decreasing:

$$1 > u_1(x) > u_2(x) > \cdots > 0 \quad \text{for all } x \in B.$$

A careful integration by parts, initially avoiding the $(n-2)$-dimensional sphere $\{x; \; |x| = R, \; x_n = 0\}$ and the boundary of 0_j, leads to

$$0 = \int_B \sum_i \frac{\partial}{\partial x_i} \left(\frac{\partial u_j / \partial x_i}{\sqrt{1 + |\mathrm{grad}\, u_j|^2}} \right) dx = \int_{\partial B} \frac{\sum_i \nu_i \frac{\partial u_j}{\partial x_i}}{\sqrt{1 + |\mathrm{grad}\, u_j|^2}} \, dS.$$

Here $\nu = (\nu_i)$ denotes the outer unit normal vector. Since $u_j(x) =$ const. almost everywhere on ∂B, we see that the tangential derivatives of $u_j(x)$ exist and are zero a.e. on ∂B, so that

$$\int_{\partial B} \frac{\partial u_j / \partial \nu}{\sqrt{1 + (\partial u_j / \partial \nu)^2}} \, dS = 0.$$

Obviously,

$$\frac{\partial u_j}{\partial \nu} > 0 \text{ on } 0_j, \qquad \frac{\partial u_j}{\partial \nu} < 0 \text{ on } B_j.$$

For $j = 1, 2, \ldots$ set

$$a_j = \int_E \frac{\partial u_j / \partial \nu}{\sqrt{1 + (\partial u_j / \partial \nu)^2}} \, dS > 0.$$

It follows from the monotone character of the $u_j(x)$ that

$$0 < a_1 < a_2 < \cdots \quad \text{and} \quad 0 < a = \lim_{j \to \infty} a_j < R^{n-1} \omega_{n-1}.$$

The solutions $u_j(x)$ converge uniformly in every compact subset of B to a limit function

$$u(x) \in C^2(B) \cap C^0(B \cup (\partial B \setminus E)).$$

This limit function is bounded, $0 \leqslant u(x) < 1$, and is itself a solution of $(*)$ in B.

Choose an arbitrary point $x_0 \in B$. *We assert that $u(x_0) > 0$. In words: If $H_{n-1}(E) > 0$, the boundary set $E \subset \partial B$ supports a nontrivial solution of* (*).

Proof. Otherwise, $u_j(x_0) \downarrow 0$. Then, in view of the gradient estimate for solutions of (*) and with the help of the maximum principle and suitably chosen barriers, it can be concluded that the convergence $u_j(x) \downarrow 0$ is uniform in each

$$B_\delta = \{x; \ |x| < R, \ x_n > \delta\}, \ \delta > 0.$$

Consider a point $\bar{x} \in S \setminus E$ and choose $\delta' > 0$ so small and j so large, say $j \geqslant j'$, that the whole neighborhood

$$N_{\delta'} = \{x; \ |x - \bar{x}| < \delta', \ x_n = 0\}$$

is contained in $S \setminus 0_j$. For sufficiently small η, the cylinder

$$C_{\delta', \eta} = N_{\delta'} \times \{|x_n| < \eta\}$$

over $N_{\delta'}$ lies in B. If $j \geqslant j'$ and $x \in N_{\delta'}$, we have $u_j(x) = 0$. $u_j(x)$ can be extended by reflection on S to a solution of the minimal surface equation in $C_{\delta', \eta}$. The gradient estimate combined with our earlier statement about the behavior of the functions $u_j(x)$ in each B_δ allows us to conclude that the extensions of the $u_j(x)$ converge to zero, uniformly in $C_{\delta', \eta}$. It follows that $\partial u_j / \partial \nu \to 0$ on $\partial B \setminus E$, the convergence being uniform on every compact subset of $\partial B \setminus E$.

Let a number $\varepsilon > 0$ be given. Choose $N = N(\varepsilon) > 0$ so large that

$$a_j > \frac{a}{2}, \quad H_{n-1}(0_j \setminus E) < \varepsilon \quad \text{for } j \geqslant N.$$

We have

$$0 = \int_{\partial B} \frac{\partial u_j / \partial \nu}{\sqrt{1 + (\partial u_j / \partial \nu)^2}} \, dS$$

$$= \int_{B_N} + \int_{B_j \setminus B_N} + \int_{0_j \setminus E} + \int_E \frac{\partial u_j / \partial \nu}{\sqrt{1 + (\partial u_j / \partial \nu)^2}} \, dS$$

$$= I_j^{(1)} + I_j^{(2)} + I_j^{(3)} + I_j^{(4)}.$$

Now,

$$\lim_{j \to \infty} I_j^{(1)} = 0, \qquad |I_j^{(2)}| \leqslant \varepsilon, \qquad |I_j^{(3)}| \leqslant \varepsilon, \qquad I_j^{(4)} = a_j > \frac{a}{2}.$$

The resulting inequality

$$I_j^{(1)} \leqslant 2\varepsilon - \frac{a}{2}$$

leads to a contradiction if ε is chosen smaller than $a/4$. The assertion, $u(x_0) > 0$, is proved

Let us extend $u(x)$ by reflection on $S \setminus E$ into the full ball $D = \{x; \ |x| < R\}$. The extended function is a C^2-solution of $(*)$ in $D \setminus E$. It is continuous in $\overline{D} \setminus E$ and vanishes on the boundary ∂D. There is of course a (unique) C^2-solution $\bar{u}(x)$ of the minimal surface equation in D with boundary values zero, namely, $\bar{u}(x) = 0$. Since $u(x) \not\equiv 0$ in D, the set E cannot be removable.

REFERENCES

1. F. J. Almgren, *Plateau's problem. An Invitation to Varifold Geometry*, Benjamin, New York-Amsterdam, 1966.
2. ———, "Existence and regularity almost everywhere of solutions to elliptic variational problems with constraints," *Mem. Amer. Math. Soc.*, vol. 4, **165** (1976).
3. ———, "Minimal surfaces: tangent cones, singularities, and topological types," *Proc. Internat. Congr. Math.*, Helsinki (1978), 767–770.
4. R. Böhme and A. J. Tromba, "The index theorem for classical minimal surfaces," *Ann. of Math.*, **113** (1981), 447–499.
5. E. Bombieri, "Recent progress in the theory of minimal surfaces," *Enseign. Math.*, **27** (1979), 35–42.
6. E. Bombieri, E. DeGiorgi, and E. Giusti, "Minimal cones and the Bernstein theorem," *Invent. Math.*, **7** (1969), 243–269.
7. M. do Carmo, "Minimal surfaces: stability and finiteness," *Proc. Internat. Congr. Math.*, Helsinki (1978), 401–405.
8. B. Y. Chen, *Geometry of Submanifolds*, Dekker, New York, 1973.
9. S. S. Chern, "Minimal surfaces in a Euclidean space of N dimensions," *Differential and Combinatorial Topology*, S. S. Cairns, ed., Princeton Univ. Press (1965), 187–198.
10. S. S. Chern and R. Osserman, "Complete minimal surfaces in Euclidean n-space," *J. d'Anal. Math.*, **19** (1967), 15–34.

11. R. Courant, *Dirichlet's Principle, Conformal Mapping, and Minimal Surfaces*, Interscience, New York, 1950.

12. H. Federer, *Geometric Measure Theory*, Springer-Verlag, New York, 1969.

13. D. Fischer-Colbrie, "Some rigidity theorems for minimal submanifolds of the sphere," *Acta Math.*, **145** (1980), 29–46.

14. D. Gilbarg and N. S. Trudinger, *Elliptic Partial Differential Equations of Second Order*, Springer-Verlag, Berlin-Heidelberg-New York, 1977.

15. R. Hardt and L. Simon, "Boundary regularity and embedded solutions for the oriented Plateau problem," *Ann. of Math.*, **110** (1979), 439–486.

16. S. Hildebrandt, "Liouville theorems for harmonic mappings and an approach to Bernstein theorems," *Ann. of Math. Studies*, to appear.

17. S. Hildebrandt, J. Jost, and K. O. Widman, "Harmonic mappings and minimal submanifolds," *Invent. Math.*, **62** (1980), 29–46.

18. D. A. Hoffman and R. Osserman, "The geometry of the generalized Gauss map," *Mem. Amer. Math. Soc.*, vol. 28, **236** (1980).

19. H. B. Lawson, *Lectures on Minimal Submanifolds*, IMPA, Rio de Janeiro, 1973. (2nd ed. reprinted by Publish or Perish Press, Berkeley, 1980.)

20. ———, *Minimal Varieties in Real and Complex Geometry*, Univ. of Montreal Press, Montreal, 1973.

21. ———, "Minimal varieties," *Proc. Symp. Pure Math.*, Amer. Math. Soc., vol. **27**, part 1 (1975), 143–175.

22. W. Meeks, *Lectures on Plateau's Problem*, IMPA, Rio de Janeiro, 1978.

23. J. C. C. Nitsche, "On new results in the theory of minimal surfaces," *Bull. Amer. Math. Soc.*, **71** (1965), 195–270.

24. ———, "A new uniqueness theorem for minimal surfaces," *Arch. Rational Mech. Anal.*, **52** (1973), 319–329.

25. ———, *Vorlesungen über Minimalflächen*, Springer-Verlag, Berlin-Heidelberg-New York, 1975.

26. ———, "Plateau's problems and their modern ramifications," *Amer. Math Monthly*, **81** (1974), 945–968.

27. ———, "The higher regularity of liquid edges in aggregates of minimal surfaces," *Nachr. Akad. Wiss. Göttingen Math.-Phys. Kl. 2* (1978), 31–51.

28. M. Obata, ed., *Minimal Submanifolds and Geodesics*, Kagai Publ., Tokyo, 1978.

29. R. Osserman, *A Survey of Minimal Surfaces*, Van Nostrand Reinhold, New York, 1969.

30. ———, "Minimal varieties," *Bull. Amer. Math. Soc.*, **75** (1969), 1092–1120.

31. T. Radó, "On the problem of Plateau," *Ergebn. d. Math. u. ihrer Grenzgeb.*, Springer-Verlag, Berlin, 1933.

32. J. Simons, "Minimal varieties in Riemannian manifolds," *Ann. of Math.*, **88** (1968), 62–105.

33. J. E. Taylor, "The structure of singularities in soap-bubble-like and soap-film-like minimal surfaces," *Ann. of Math.*, **103** (1976), 489–539.

34. F. Tomi, "On the local uniqueness of the problem of least area," *Arch. Rational Mech. Anal.*, **52** (1973), 312–318.

January 1980

PROBABILISTIC METHODS IN
PARTIAL DIFFERENTIAL EQUATIONS

*Steven Orey**

0. INTRODUCTION

Since the middle of this century, the use of probabilistic methods in partial differential equations has enjoyed considerable popularity. The purpose of this paper is to introduce these ideas to an audience not assumed to be familiar with probability, and to give some idea of the scope of these methods by presenting specific applications. In Subsection (A) these aims will be amplified and the organization of the paper explained, whilst a discrete example will be discussed in Subsection (B).

(A) Organization. Many of the applications deal with the Laplacian Δ or more general second order elliptic operators L. A function u will be called harmonic if it satisfies $Lu = 0$. In Section 2 harmonic functions will be discussed. Subsection (A) deals with the behavior of such functions near infinity, while (B) is concerned with

*Supported by NSF MCS 78-01168.

the classical Dirichlet problem and tests for the regularity of boundary points. While the purpose is expository throughout, a number of results and proofs in that section are believed to be new. In Section 3 three applications are discussed which rely on the theory of diffusion, and in particular exploit the geometry of the trajectories. In both Sections 2 and 3 the applications are to second-order linear elliptic or parabolic operators. Section 4 is devoted to showing applications involving other operators. In particular, Subsection 4(A) deals with random evolutions, and Subsection 4(B) shows how branching diffusions lead to the treatment of certain problems involving nonlinear diffusion operators.

These remarks are very brief because it seems best to let the illustrations speak for themselves and because each section has a short introduction of its own. In Section 1 a quick discussion of the relevant concepts and results needed from probability theory is given. All of the material is standard, but in order to realize the goal of making the discussion accessible to people with a minimal knowledge of probability it appeared necessary to explain the background material succinctly, with references to the textbook literature.

By now there is a huge literature dealing with the subject matter of this paper, including many hundred research papers, and numerous monographs and survey papers. No effort is made here to assign credit, and no attempt to cite original sources. Instead, an effort is made to guide the reader to expositions which give easy access to the subject or to relevant texts. Of course, when reporting on results in the research literature which have not been absorbed in the expository literature, the original source is given.

The paper is divided into Sections 0 through 4. Some sections have subsections labeled by capital Roman letters. Numbering of formulas begins anew in each subsection. A reference to "(3)" means formula (3) of the subsection of the reference; on the other hand "(1.3)" would refer to formula (3) of Section 1, etc.

(B) A Discrete Problem. Before explaining the connection between certain differential operators and corresponding Markov processes, it will be easier to illustrate matters in a discrete situation, where the differential operators become difference operators, the Markov processes become Markov chains. Historically, it was this simpler setup which was first discussed.

Let E^2 be the Euclidean lattice of points (m, n), with m and n integers. If u is a real valued function defined on E^2, let

$$Lu(m, n) = \tfrac{1}{4}[u(m+1, n) + u(m-1, n)$$
$$+ u(m, n+1) + u(n, n-1)] - u(m, n).$$

Let $A \subseteq E^2$ be the rectangle defined by

$$A = \{(m, n): a < m < b, c < n < d\}.$$

Let

$$\partial A = \{(m, n): c < n < d; m \in \{a, b\}\}$$
$$\cup \{(m, n): a < m < b, n \in \{c, d\}\}.$$

Now consider the discrete Dirichlet problem

$$Lu(m, n) = 0, \quad (m, n) \in A, \quad u(m, n) = \phi(m, n), \quad (m, n) \in \partial A.$$
$$\tag{1}$$

Let X_t, $t = 0, 1, \ldots$ be the position of a random walk on E^2 which, at each instant of time, changes position by moving from its present position to one of the four neighboring positions, choosing these positions with equal probabilities $1/4$ each. To calculate probabilities related to the moving particle it is still necessary to specify the point where the particle starts at time $t = 0$, and so the dependence of the probability measure P will be indicated by a superscript. For example $P^{(1,2)}[X_1 = (1,1)] = 1/4$. Let T be the least $t \geq 0$ such that $X_t \in \partial A$. Then $T = 0$ if $X_0 \in \partial A$, and $0 < T < \infty$ holds with $P^{(x, y)}$-probability one if $X_0 = (x, y) \in A$. The solution to (1) can be represented as

$$u(m, n) = \sum_{(p, q) \in \partial A} \phi(p, q) P^{(m, n)}[X_T = (p, q)],$$

$$(m, n) \in A \cup \partial A. \tag{2}$$

Since for $(m, n) \in \partial A$, $P^{(m, n)}[T = 0] = 1$, the boundary condition obviously holds. Next let $(m, n) \in A$ and let $N(m, n)$ be the four neighbors of (m, n). Observe that (2) implies

$$u(m, n) = \sum_{(r, s) \in N(m, n)} \sum_{(p, q) \in \partial A} \phi(p, q) P^{(m, n)}[X_1 = (r, s), X_T = (p, q)]$$

$$= \frac{1}{4} \sum_{(r, s) \in N(m, n)} u(r, s) = [(L + I)u](m, n)$$

where I is the identity. So $Lu(m, n) = 0$.

In the setup just described one is dealing not with a single probability measure, but with a family $P^{(m, n)}$ indexed by the starting point. The actual probability space is not very relevant, but frequently it is convenient to take the space to be the space of all random walk paths, that is, sequences $\omega = (\omega_i)$, $i = 0, 1, \ldots$, and to let $X_k(\omega) = \omega_k$. Then

$$P^{(m, n)}\{\omega: \omega_0 = (m_0, n_0), \ldots, \omega_k = (m_k, n_k)\}$$

$$= P^{(m, n)}[X_0 = (m_0, n_0), \ldots, X_k = (m_k, n_k)],$$

and this quantity equals 4^{-k} if $(m_0, n_0) = (m, n)$ and (m_{i+1}, n_{i+1}) is a neighbor of (m_i, n_i), $i = 1, 2, \ldots, k$; otherwise it equals zero. So the random walk is given by specifying the collection of probability measures $P^{(m, n)}$. Sometimes an alternative point of view is more convenient: one thinks of the random walk as a collection of processes $(X_t^{(m, n)})$, $(m, n) \in E^2$, all defined on the same probability space, with one probability measure P. Indeed, in the present very special example, one can define all the processes in terms of $X^{(0,0)}$ by $X_t^{(m, n)} = X_t^{(0,0)} + (m, n)$. Now $P[X_0^{(m, n)} = (m, n)] = 1$, and as before,

$$P\left[X_{t+1}^{(m, n)} = (r, s) \mid X_t^{(m, n)} = (m_k, n_k), k = 0, 1, \ldots, t\right] = \tfrac{1}{4}$$

provided (r, s) is a neighbor of (m_t, n_t). Both these notational approaches are in common use, and both will be used below. If E, $E^{(m, n)}$ are the expectation operators corresponding to P, $P^{(m, n)}$, respectively, the representation (2) can be written as follows, using the original approach

$$u(m, n) = E^{(m, n)}\big[\phi(X_T)\big].$$

Under the second approach this would take the form

$$u(m, n) = E\big[\phi(X_{T^{(m, n)}}^{(m, n)})\big].$$

Evidently one can generalize the above example by taking more general random walks in which the transitions are not equally likely to the nearest neighbors. However discrete problems will not be pursued here. For an excellent introduction see Dynkin and Yushkevitch [12].

1. THE TOOLS OF PROBABILITY

The main ideas and results from probability theory which will be needed in the applications of the subsequent sections are introduced and explained briefly. By now there is an extensive textbook and monograph literature. This section does not compete with it. Rather it is hoped that it provides the basic definitions and makes plausible the main theorems so that the uninitiated reader can proceed through the rest of the paper. For those wishing to master the details numerous references are provided.[1]

(a) Markov Processes and Their Generators. Most commonly the connection between partial differential equations and probability are due to an intimate relationship between a differential operator L and the generator of a Markov process; in many cases L actually is the differential generator. The study of diffusions in R^d leads to second order elliptic operators. An operator of the form

$$L = \frac{1}{2} \sum_{i, j = 1}^{d} a_{ij}(x) \frac{\partial^2}{\partial x_i \partial x_j} + \sum_{i = 1}^{d} b_i(x) \frac{\partial}{\partial x_i} \tag{1}$$

---

[1]For a treatment of stochastic calculus and many applications the recent monograph, *Stochastic Differential Equations and Diffusion Processes*, by N. Ikeda and S. Watanabe, North-Holland, 1981, is highly recommended.

will be called a *canonical diffusion operator* if the coefficients $a_{ij}(x)$ and $b_i(x)$ are uniformly bounded, uniformly Hölder continuous, and (a_{ij}) is uniformly positive definite, which is to say, there exists $\lambda_0 > 0$ such that

$$\sum_{i,j=1}^{d} a_{ij}(x)\xi_i\xi_j \geqslant \lambda_0|\xi|^2, \quad x \in R^d, \quad \xi \in R^d.$$

Some background from the theory of Markov processes must be reviewed; for a basic reference see [11]. Analytically a *time-homogeneous Markov process* on a measurable space $(S, \mathcal{B})$ is determined by its transition probability function $p(t, x, B)$ which satisfies the following conditions: For fixed t and x, $p(t, x, \cdot)$ is a probability measure on $\mathcal{B}$; for fixed t and B, $p(t, \cdot, B)$ is $\mathcal{B}$-measurable; and

$$p(s + t, x, B) = \int_S p(s, x, dy)p(t, y, B).$$

On the Banach space of all bounded measurable functions, define operators T_t by

$$T_t f(x) = \int p(t, x, dy)f(y), \quad t \geqslant 0. \tag{2}$$

Then evidently the T_t form a semigroup $(T_{t+s} = T_t \cdot T_s)$ with $T_0 = I$, the identity operator. Knowing the T_t is equivalent to knowing the transition probabilities. The semigroup in turn is known to be completely determined by its infinitesimal generator $\mathcal{L}$ defined by

$$\mathcal{L}f = \lim_{t \to 0} \frac{1}{t}(T_t - I)f \tag{3}$$

where the limit is in the supremum norm, and the domain of $\mathcal{L}$ consists of all f for which the limit exists. The operator $\mathcal{L}$ is linear, but usually unbounded. For details one may consult Dynkin [11] or the excellent exposition in Loève [26]. An easy but useful consequence of the semigroup property is that for f in the domain of $\mathcal{L}$

$$\frac{d}{dt}[T_t f] = \mathcal{L}T_t f. \tag{4}$$

For given transition probabilities $p(t, x, B)$ there exists a corresponding Markov process X_t. That is, for $0 \leq t < \infty$ there exist random variables X_t on a probability space $(\Omega, \mathcal{F})$ with values in $(S, \mathcal{B})$ and, for each $x \in S$, there is a probability measure P^x on Ω corresponding to "the process started at x" and satisfying: for $0 \leq t_1 < t_2 < \cdots < t_n$ and $B_1, B_2, \ldots, B_n \in \mathcal{B}$,

$$P^x \left[X_{t_1} \in B_1, X_{t_2} \in B_2, \ldots, X_{t_n} \in B_n \right]$$

$$= \int_{B_n} \cdots \int_{B_2} \int_{B_1} p(t_1, x, dx_1)$$

$$\times p(t_2 - t_1, x_1, dx_2) \cdots p(t_n - t_{n-1}, x_{n-1}, dx_n).$$

$$(5)$$

It is clear that if $\mathcal{F}_\infty$ denotes the smallest σ-field including all events of the form appearing on the left side of (5), P^x is determined on $\mathcal{F}_\infty$. Events of the above form, but with t_n restricted to be less than or equal to a fixed t, generate a smaller σ-field $\mathcal{F}_t$. Intuitively it consists of all events whose occurrence or nonoccurrence is determined by time t. It is necessary to consider conditional probabilities of the form $P^x[\Lambda | \mathcal{F}_t]$. The reader not familiar with this notation may interpret it as the conditional P^x-probability of Λ, given the entire history of the process X_s up to time t, i.e., $P^x[\Lambda | X_s, 0 \leq s \leq t]$. The *Markov property* connects the probability measures P^x as follows: letting $g(x)$ stand for the function on the left side of (5),

$$P^x \left[X_{t+t_1} \in B_1, X_{t+t_2} \in B_2, \ldots, X_{t+t_n} \in B_n | \mathcal{F}_t \right] = g(X_t). \quad (6)$$

Often a stronger form of this property is required. Call a positive random variable T which is $\mathcal{F}_\infty$-measurable a *stopping time* if $[T < t] \in \mathcal{F}_t$. Note that here and throughout notations such as $[T < t]$ stand for events such as $\{\omega \in \Omega : T(\omega) < t\}$. Let

$$\mathcal{F}_{T+} = \{ \Lambda \in \mathcal{F}_\infty : \text{for each positive } t, \Lambda \cap [T < t] \in \mathcal{F}_t \}.$$

Intuitively, T is a stopping time if one can determine that $T = t$ by observing the process X_s for $0 \leqslant s < t + \varepsilon$, where $\varepsilon > 0$ can be arbitrarily small. Requiring $\varepsilon = 0$ would result in a more restrictive notion of stopping time. The *strong Markov property* holds if for every stopping time T

$$P^x\Big[X_{T+t_1} \in B_1, X_{T+t_2} \in B_2, \ldots, X_{T+t_n} \in B_n | \mathcal{F}_{T+} \Big] = g(X_T) \quad (7)$$

with g as in (6). There are two aspects making (7) stronger than (5): T need not be a constant, and even in case $T \equiv t_0$, $\mathcal{F}_{T+}$ is bigger than $\mathcal{F}_{t_0}$. Though examples of Markov processes that are not strong Markov processes can be contrived, under mild assumptions the strong Markov property can be proved to hold. Using this it is not difficult to prove the Blumenthal *zero-one law* which asserts that for any $x \in S$, if $T \equiv 0$ and $\Lambda \in \mathcal{F}_{T+}$, then $P^x[\Lambda]$ must equal zero or one. The hypothesis means that Λ is an event which depends on knowing what happens after time zero for only arbitrarily short time periods.

A particularly simple kind of Markov process is a *jump process* on a finite set S; here $\mathcal{B}$ will be taken to be the class of all subsets of S. Starting at x, the process remains there for a random time τ, then moves to a new position y, chosen at random, after which the procedure repeats. So the sample functions are step functions. In this case the differential generator is completely determined by knowing

$$\lim_{t \to 0} \frac{p(t, x, \{y\}) - 1}{t} \equiv q(x, y), \quad x, y \in S.$$

Writing $q(x)$ for $-q(x, x)$, it is easy to show that τ has an exponential distribution

$$P^x[\tau > t] = e^{-q(x)t}.$$

Thus

$$P^x[\text{jump during } [0, t]] = q(x)t + o(t)$$

and, for $t \to 0$,

$$P^x[\text{jump to } y \text{ during } [0, t] | \text{some jump during } [0, t]] \to \frac{q(x, y)}{q(x)}.$$

This is the simplest kind of time homogeneous Markov process. It will be encountered in Section 4(A). Jump processes on a denumerable set S are also frequently encountered. For example the *Poisson process with intensity a* resides on $S = \{0, 1, \ldots\}$ and is determined by $q(n, n + 1) = a, q(n, n) = -a$ for $n = 0, 1, \ldots, q(n, m) = 0$ in all other cases.

A homogeneous Markov process with the strong Markov property taking values in $(R^d, \mathfrak{B}^d)$, with $\mathfrak{B}^d$ the d-dimensional Borel sets, will be called a *diffusion process* provided the *sample functions*, $t \to X_t$ are continuous P^x-a.s. (read "almost surely", i.e., with P^x-probability one) for each x. This means

$$P^x\{\omega: t \to X_t(\omega) \text{ is continuous on } 0 \leqslant t < \infty\} = 1.$$

So in some sense these are opposites of the jump processes introduced above.

The discussion will now be specialized to diffusion processes, even though most of it can be done in general, see Dynkin [11], Loève [26], Blumenthal and Getoor [4].

(b) Diffusion Processes. Let $X = (X_t)$ be a diffusion process. The notation

$$T_A = \inf\{t: t > 0, X_t \in A\} \tag{8}$$

will be used throughout. If A is either open or closed, T_A can easily be shown to be a random variable; results for the case that A is only assumed to be a Borel set are more difficult to establish. Given that T_A is a random variable, it is easily seen to be a stopping time. A^c is used to denote the complement of A.

The Dynkin *generator* $\bar{\mathbb{L}}$ is a fruitful device. The idea is to replace the limit $t \to 0$ in (3) by a limit along a suitable class of stopping times. For f a bounded measurable function on R^d, $x \in R^d$, let

$$\bar{\mathbb{L}}f(x) = g(x)$$

if and only if

$$g(x) = \lim_{U \downarrow x} \frac{E^x\left[f\left(X_{T_{U^c}}\right)\right] - f(x)}{E^x[T_{U^c}]}$$

where U ranges over neighborhoods of x, $U \downarrow x$ indicating that the diameter of U approaches zero. Under mild assumptions it can be shown that $\mathcal{L}f = g$ implies that for each x, $\bar{\mathcal{L}}f(x) = g(x)$. For a complete discussion see [11]; for elementary examples see [26].

If X is a diffusion with Dynkin generator $\bar{\mathcal{L}}$ such that for each $x \in R^d$, every bounded measurable function f which is in C^2 in some neighborhood of x, satisfies

$$\bar{\mathcal{L}}f(x) = Lf(x)$$

where L is a differential operator, then L will be said to be the *differential generator* of X; conversely X is the *characteristic process* for the operator L. The characteristic process of $\frac{1}{2}\Delta$ is known as the *Wiener process* or as *Brownian motion*. The infinitesimal generator $\mathcal{L}$ defined by (3) is an extension of L. Given a canonical diffusion operator L, it determines $\mathcal{L}$; see [11], [26]. The terminology "characteristic process" is chosen for the following reason: if L is of the form (1) with all $(a_{ij}) \equiv 0$, so that L is a first order operator, the corresponding characteristic process degenerates to a deterministic process. That is, P^x puts total mass on the path X_t which starts at x and satisfies

$$X_t = x + \int_0^t b(X_s)\, ds. \tag{9}$$

This of course is just the solution $X_t^{(x)}$ of the differential equation

$$dX_t^{(x)} = b\left(X_t^{(x)}\right) dt, \qquad X_0^{(x)} = x. \tag{10}$$

These are just the usual characteristic curves associated with the first order operator L in this case. It will be explained below how a generalization of (10) leads to the theory of stochastic differential equations.

If L is a canonical diffusion operator, the characteristic process will be called a *canonical diffusion*. Such processes are known to enjoy numerous nice properties.

To begin with one has *transition probability densities* so that, for $t > 0$,

$$p(t, x, B) = \int_B p(t, x, y)\, dy, \quad x \in R^d, \quad B \in \mathfrak{B},$$

where

$$p(t, x, y) = p(s, x; s + t, y)$$

with $p(s, x; t, y)$ being the fundamental solution corresponding to the operator $\partial/\partial t + L$. Such fundamental solutions are known to have certain regularity properties, see [11] or [13]. In particular $p(s, x, y)$ is continuous in x. This implies immediately the so-called *strong Feller property*: for f bounded and measurable, $t > 0$, the function $T_t f$ defined by (2) is continuous. In turn it follows easily that if U is open and ϕ is a bounded measurable function defined on ∂U, then the function

$$h(x) \equiv E^x\big[\phi(X_{T_{\partial U}})\big] \equiv \int_{T_{\partial U} < \infty} \phi(X_{T_{\partial U}})\, dP^x \tag{11}$$

is continuous in U. Canonical diffusion operators are well known to satisfy the *strong maximum principle*: if D is an open connected set and u is a $C^2(D)$-function such that $Lu \geq 0$ in D, then u cannot attain a positive maximum in D, unless $u \equiv$ constant in D.

Given a diffusion X it is sometimes useful to consider a topology intrinsically associated with X, the *fine topology*: a set U is *finely open* if for every $x \in U$, P^x-a.s. the path X_t remains within U for a positive length of time. Since the paths are continuous, evidently every set which is open in the Euclidean topology is finely open.

Let X be a diffusion in R^d. Then the process $\tilde{X}$ given by $\tilde{X}_t = (X_t, Z_t)$ where Z_t is a deterministic process moving in R^1 with constant velocity one (i.e., $Z_t = Z_0 + t$) is a diffusion in R^{d+1}. If X has a differential generator L, $\tilde{X}$ has differential generator $\tilde{L}$ given by

$$\tilde{L}f(x,t) = L_x f(x,t) + \frac{\partial f}{\partial t}(x,t).$$

The process $\tilde{X}$ is the *space-time process* associated with X. If X is a canonical diffusion, $\tilde{X}$ will be called a *canonical space-time process*. Evidently $\tilde{X}$ is *not* a canonical diffusion, since $\tilde{L}$ is parabolic. All information contained in X will also be contained in $\tilde{X}$. Incidentally, one is led to a method of dealing with diffusion operators L which differ from the canonical diffusion operator in that the coefficients depend on t. There do in fact exist corresponding Markov processes X, but they have transition probabilities of the form $p(x, s, dy, t)$; that is, the conditional probability that $X_t \in dy$ given that $X_s = x$, where $s < t$, depends on s and t, not just on $t - s$. While it is quite possible to work with such nonhomogeneous transition probabilities, it can be avoided. For the associated space-time process $\tilde{X}$ still has homogeneous transition probabilities. For B, a suitable Borel set in $R^d \times R^1$, the transition probabilities clearly are given by

$$\tilde{p}(t,(x,s), B) = p(x, s; B_t, t) = \int_{B_t} p(x, s; y, t)\, dy, \quad (12)$$

where B_t is the section of B determined by t, $t \geqslant s_0$.

If $\tilde{X}$ is a canonical space time process and $\tilde{T}_t$ the associated semigroup, it is still the case that $\tilde{T}_t f$ is continuous provided f is bounded and continuous; this is known as the *Feller property*.

(c) Potential Theory. The classical potential theory associated with the Laplacian can be extended to the differential generators of diffusions, and even to generators of quite general Markov processes. Not only does the analytical theory extend, but many concepts have probabilistic interpretations in terms of the characteristic process. This provides one of the main links between probabilistic and analytic problems. For elementary introductions to this theory see [12], [33], [31], [32], and for more comprehensive treatments [4], [11], [13]; for parabolic problems [9] is the prime reference. Here the aim is the very limited one of explaining some of the relevant concepts which will be needed in the applications in subsequent sections.

Let L be the differential generator of a diffusion X. For U an open set, recall that

$$T_{U^c} = \inf\{t > 0: X_t \in U^c\}$$

denotes the *first exit time* from U. Let G be an open set. A measurable function f from G to the reals will be said to have the *mean value property* if it is bounded on compact subsets of G, and for every $x \in G$ and any open set U with compact closure contained in G,

$$f(x) = E^x\left[f\left(X_{T_{U^c}}\right)\right]. \tag{13}$$

Now let L be a canonical diffusion operator, x the characteristic process in R^d. Then one has the

Criterion for harmonic functions: For G an open subset of R^d, f a measurable function on G, bounded on compact subsets of G, the following are equivalent:

(i) $f \in C^2(G)$ and $Lf(x) = 0$ for all $x \in G$,
(ii) f is continuous in G and $\bar{\mathcal{L}}f(x) = 0$ for all $x \in G$,
(iii) f has the mean value property on G.

For details see Dynkin [11], Chapter 12, and Friedman [13]. The function f on G will be called superharmonic on G if it is lower semicontinuous, bounded below on compact subsets of G, and has the *supermean value* property obtained by replacing $=$ by $\geqslant$ in (13). Again see [11] for further information. If f satisfies the equivalent conditions (i)–(iii) above, it will be called *harmonic* on G.

If X is a canonical diffusion, U an open set with compact closure, then $T_{U^c} < \infty$, P^x-a.s., *and even* $E^x[T_{U^c}] < \infty$; for an easy proof see Friedman [13], Theorem 5.1. This result uses the ellipticity of (a_{ij}).

Let G be an open set with compact closure, ϕ a bounded measurable function defined on ∂G, and define

$$h(x) = E^x[\phi(X_T)], \qquad T = T_{G^c}. \tag{14}$$

Then h gives a "solution" to the Dirichlet problem

$$Lu = 0 \text{ in } G, \qquad u = \phi \text{ on } \partial G.$$

For if U is an open set with compact closure in G, $x \in U$, the process started at x must leave U before leaving G, and, using the strong Markov property,

$$h(x) = \int h(y) P^x[X_{T_{\partial U}} \in dy]$$

so that h has the mean value property. Of course without some additional assumptions one cannot expect $h(y) \to \phi(z)$ as $y \to z$ through G. However if ∂G is smooth and ϕ is continuous, then for $y \in G$ near $z \in \partial G$, one would expect that the process started at y leaves G near z, hence $h(y)$ will be close to $\phi(z)$. Even without further assumptions it can be shown that

$$\lim_{t \uparrow T} h(X_t) = \phi(X_T), P^x\text{-a.s.}, x \in G.$$

See Dynkin [11], Theorem 12.12.

A canonical diffusion X with transition probability function $p(t, x, A)$ will be said to have a *potential function g* if

$$g(x, A) = \int_0^\infty p(t, x, A) \, dt < \infty, x \in R^d, A \text{ compact.}$$

This is the case for Brownian motion in dimension $d > 2$, but not for $d = 1, 2$. In case the potential fails to exist one can work just about as well with the λ-*potential*

$$g_\lambda(x, A) = \int_0^\infty e^{-\lambda t} p(t, x, A) \, dt, \qquad \lambda > 0.$$

For simplicity only, the discussion will be restricted now to the case that the potential function exists. If then X is a canonical diffusion with potential function g, one has

$$g(x, dy) = g(x, y)\, dy = \int_0^\infty p(t, x, y)\, dt\, dy$$

and $g(x, y)$ is the *potential kernel*. For Brownian motion it reduces to the usual Newtonian kernel. A positive measure on μ on R^d has a *potential* given by

$$f(x) = \int g(x, y)\mu(dy).$$

For a Borel set B define

$$h_B(x) = P^x[X_t \in B \text{ for some } t > 0].$$

If B has compact closure, h_B can be shown to be the potential of a certain measure μ_B concentrated on B^{closure}, the so-called equilibrium potential. For details consult Blumenthal and Getoor [4] and Dynkin [11]. For introduction to the theory in the case of Brownian motion, M. Rao [33] and Port and Stone [31] and [32] are recommended.

All the above-mentioned results have analogues for canonical space-time diffusions and their infinitesimal generators ($\partial/\partial t + L$). This applies in particular to the discussion of the Dirichlet problem. For the space time process the potential kernel is naturally taken to be

$$\tilde{g}((x, s), (y, t)) = p(t - s, x, y), \qquad t \geqslant s.$$

The theory is developed extensively in [9].

(d) Generalizations. The probabilistic framework can be extended in many different directions to deal with operators L more general than, or quite different from, the canonical diffusion operator.

To begin with, one obviously need not restrict oneself to Euclidean space. Often one wants to consider diffusions on subsets of Euclidean space; this is necessary for a treatment of the Neumann problem. Frequently and naturally then, the theory is developed on manifolds.

Considering diffusion operators in R^d one may wish to relax the conditions imposed in the definition of a canonical diffusion operator. There are two quite different sorts of weakenings of assumptions. If the assumption of ellipticity is dropped, and (a_{ij}) is assumed to be only nonnegative definite, one can, at least under more stringent continuity assumptions on the coefficients, still obtain a characteristic process and use it for investigating problems in differential equations. Of course some of the properties noted above—the strong maximum principle for example—will now fail. Nevertheless this has proved a fruitful way of investigating degenerate elliptic and parabolic p.d.e., see Friedman [13]. An important application, the Stroock-Varadhan support theorem, will be given in Section 3(B). On the other hand, one may retain the assumption of uniform ellipticity but assume that the coefficients are only continuous. It turns out that the relevant theorems become vastly more difficult to establish—beginning with the existence of the characteristic process. However, thanks to the work of Stroock and Varadhan (see [38]) it is known that these processes not only exist, but enjoy most of the properties of canonical diffusions.

One can also deal with operators L which have a quite different form from diffusion operators. In Section 4(A) probabilistic treatment of certain hyperbolic operators will be briefly discussed. In Section 4(B) it will be indicated how operators arising from canonical diffusion operators by addition of a nonlinear term have been dealt with by using branching diffusions.

(e) Martingales. Let X be a canonical diffusion in R^d, and let $\mathcal{F}_t, P^x$, be as in (a). Sometimes one has occasion to deal with a real-valued process $H = (H_t)$ which has the properties that for each t, H_t is $\mathcal{F}_t$ measurable and, for fixed x,

$$E^x[H_t|\mathcal{F}_s] = H_s, \qquad 0 \leqslant s \leqslant t < \infty. \tag{15}$$

The above says that, with respect to the probability measure P^x, $(H_t, \mathfrak{F}_t, t \geq 0)$ constitutes a *martingale* process. Martingales have been extensively studied and the basic theory can be found in many books, for example Loève [26]. No attempt to summarize the results will be made here. Some relevant examples should be noted. If h is a harmonic function, $H_t = h(X_t)$ will be a martingale for each P^x. If furthermore $h \geq 0$, the famous martingale convergence theorem implies

$$\lim_{t \to \infty} H_t = H_\infty P^x\text{-a.s.}$$

For a martingale one certainly has

$$E^x[H_t] = E^x[H_0], \quad 0 \leq t < \infty. \tag{16}$$

In case $E^x|H_t|$ is bounded uniformly in t

$$E^x[H_0] = E[H_\infty], \tag{17}$$

and if $H_t = h(X_t)$ the left side of (17) is $h(x)$.

A second important example is obtained by taking $\Lambda \in \mathfrak{F}_\infty$ and letting, for fixed x,

$$M_t = E^x[\Lambda | \mathfrak{F}_t].$$

Then $M_t \to M_\infty$ and outside a P^x-null set M_∞ equals one on Λ and zero on Λ^c. This important result, which goes back to Paul Lévy, says that for an event $\Lambda \in \mathfrak{F}_\infty$ one can predict with an arbitrarily high degree of accuracy whether or not Λ happens by watching long enough.

The relation (16) can fail if t is replaced by a stopping time T. A basic result of Doob however says that such a replacement is justified under suitable conditions on T; the boundedness of T is a sufficient condition.

If $\geq$ replaces $=$ in (15), one obtains a *supermartingale*, rather than a martingale. Many of the basic properties of martingales extend to supermartingales. In the context of diffusion processes superharmonic functions lead to supermartingales just as harmonic functions produce martingales.

(f) Stochastic Integrals. The Wiener process above was introduced as the characteristic process corresponding to the differential generator $\frac{1}{2}\Delta$. Here the emphasis will be different. In particular at present there is no necessity to start the process at different points. So instead of dealing with a family P^x of probability measures only $P \equiv P^0$, with 0 being the origin, will be needed. So the process, now to be denoted by $W = (W_t)$, will satisfy $P[W_0 = 0] = 1$. The values of W_t lie in R^d.

Consider first the case $d = 1$. An interpretation is to be given to integrals of the form

$$Y_t = \int_0^t H_s \, dW_s \tag{18}$$

where H_s may itself be random, e.g., $H_s = s + W_s$. Here s is the variable of integration. If, as a function of s, W_s were a.s. of bounded variation, one could simply interpret the integral as a Stieltjes integral. In fact, however, W is a.s. of unbounded variation over every parameter interval, so an alternative interpretation is required.

There is an underlying probability space $(\Omega, \mathcal{F}, P)$. In addition consider a family $\mathcal{F}_s$, $0 \leqslant s < \infty$ of sub-σ-fields of $\mathcal{F}$ satisfying (i) $\mathcal{F}_s \subseteq \mathcal{F}_t$, for $s \leqslant t$, (ii) for each s, W_s is $\mathcal{F}_s$-measurable, (iii) $E[W_t | \mathcal{F}_s] = W_s$ for $t \geqslant s$. So $(W_t, \mathcal{F}_t)$ is to be a martingale. (From elementary properties of the Wiener process it follows that taking, for example, $\mathcal{F}_t$ to be the least σ-field with respect to which W_s is measurable for $0 \leqslant s \leqslant t$ gives such a family.) Let $\mathcal{F}_\infty$ be the smallest σ-field containing all the $\mathcal{F}_t$, $0 \leqslant t < \infty$. The integrand process (H_s) will be assumed to satisfy the following conditions, t' now being a fixed positive number: (a) H_s is $\mathcal{F}_s$-measurable, $0 \leqslant s \leqslant t'$, (b) as a function of ω and s, $0 \leqslant s \leqslant t'$, $\omega \in \Omega$, $H_s(\omega)$ is measurable on $\mathcal{F}_{t'} \times$ Borel sets on $[0, t']$, (c) $E \int_0^{t'} H_s^2 \, ds < \infty$.

A process (H_s) satisfying (a), (b), and (c) is a *step-function* if there exists a partition $0 = t_0 < t_1 < \cdots < t_n = t'$ of $[0, t']$ such that $H_s = H_{t_{i-1}}$ for $t_{i-1} \leqslant s < t_i$, $i = 1, 2, \ldots, n$. For such a process the relation

$$\int_0^{t'} H_s \, dW_s = \sum_{i=1}^n H_{t_{i-1}} \left(W_{t_i} - W_{t_{i-1}} \right) \tag{19}$$

would seem to be satisfied by any reasonable notion of integral. It is now taken as a definition. The key to Ito's theory of stochastic integration is the easily checked consequence

$$E \left| \int_0^{t'} H_s \, dW_s \right|^2 = \int_0^{t'} E(|H_s|^2) \, ds. \tag{20}$$

On the left one has the square of the $L_2(P)$-norm of a random variable. If (H_s), $0 \leqslant s \leqslant t'$ is an arbitrary process satisfying (a), (b), and (c), one can find stepfunctions $H_s^{(n)}$ such that $\int_0^{t'} E|H_s^{(n)} - H_s| \, ds \to 0$ as $n \to \infty$, and it is natural to define $\int_0^{t'} H_s \, dW_s$ as the $L_2(P)$-limit of $\int_0^t H_s^{(n)} \, dW_s$. It is elementary to verify that the limit indeed exists, that it is independent of the approximating sequence, and that with this definition of integral (20) is preserved.

For $0 \leqslant t \leqslant t'$ one now defines

$$Y_t \equiv \int_0^t H_s \, dW_s = \int_0^{t'} H_s \chi_{[0, t]}(s) \, ds. \tag{21}$$

It can be shown that the resulting process Y_t, $0 \leqslant t \leqslant t'$ is a.s. continuous, and indeed $(Y_t, \mathcal{F}_t)$ is a continuous martingale. Here t' is a finite constant, but it is now easily seen that one can define

$$Y_t = \int_0^t H_s \, dW_s, \qquad 0 \leqslant t < \infty$$

provided the conditions (a), (b), (c) are satisfied for arbitrarily big finite t'.

Suppose now that the integrand process (H_s) is a.s. continuous as a function of s. Then, on an interval $[0, t']$, H can be approximated by step-functions $H^{(n)}$ as above, but now with the additional condition that the constant value of $H^{(n)}$ on $[t_{i-1}, t_i)$ is $H_{t_{i-1}}$. So in this case

$$\int_0^{t'} H_s \, dW_s = \lim_{\Delta_n \to 0} \sum_{i=1}^n H_{t_{i-1}} (W_{t_i} - W_{t_{i-1}}), \qquad \Delta_n = \max_{1 \leqslant i \leqslant n} (t_i - t_{i-1})$$

where the convergence is in $L_2(P)$. In contrast, the Stratonovitch approach to stochastic integrals is based on the limit

$$\lim_{\Delta_n \to 0} \sum_{i=1}^{n} \frac{1}{2} \left(H_{t_i} - H_{t_{i-1}} \right) \left(W_{t_i} - W_{t_{i-1}} \right).$$

This limit will exist in $L_2(dP)$ provided H satisfies (a)–(c) and, furthermore, H can be represented as

(d) $$H_s = M_s + V_s$$

where $M = (M_s)$ and $V = (V_s)$ are continuous processes satisfying (a) through (c), $(M_s, \mathcal{F}_t)$ is a martingale, and V_s, $0 \leqslant s \leqslant t'$ is a.s. of bounded variation. This leads to the Stratonovitch integral, to be denoted by $\int_0^t H_s \circ dW_s$ to distinguish it from the Ito integral defined earlier. The Stratonovitch integral exists for a more restricted class of integrals than the Ito integral. When both exist they can give different results. However, the relationship between the two is well understood. Although any statement involving the Stratonovitch integral can be translated into one involving the Ito integral, certain theorems have a simpler and more natural formulation using Stratonovitch's integral.

Integration theory for the case of the R^d-valued Wiener process is easily reduced to the one-dimensional case. If the integrand process H_s is $(m \times d)$-matrix valued, then

$$Y_t = \int_0^t H_s \, dW_s$$

is the R^m-valued process whose ith component is given by

$$Y_t^i = \sum_{k=1}^{d} \int_0^t H_s^{ik} \, dW_s^k, \qquad i = 1, 2, \ldots, m$$

where of course the individual components must satisfy the conditions discussed above. If (K_s) is R^m-valued, it is customary to use the differential notation

$$dY_t = H_t \, dW_t + K_t \, dt \qquad (22)$$

to mean

$$Y_t - Y_0 = \int_0^t H_s \, dW_s + \int_0^t K_s \, ds.$$

The notation

$$dY_t = H_t \circ dW_t + K_t \, dt \tag{23}$$

has an analogous interpretation involving Stratonovitch integrals.

Of course the vector valued case can also be treated directly, instead of working with components. See for example Arnold [1], where a very readable introduction to the topics of this and the next subsection can be found. The monographs [11] and [13] also develop the theory of Ito integration and stochastic differential equations. An enlightening discussion of the relationships between the Ito and Stratonovitch integrals is given in Ito [20].

Continuing with the vector valued case, one can see the difference between the Ito and Stratonovitch integrals from the basic change of variables formulas about to be given. Let $f: R^m \to R^1$ have two continuous bounded derivatives, and let $Z_t = f(Y_t)]$. Then Ito's *formula* asserts that (22) implies

$$dZ_t = \sum_{i=1}^m \left[\sum_{k=1}^d \frac{\partial f}{\partial y_i}(Y_t) H_t^{ik} \, dW_t^k + \frac{\partial f}{\partial y_i}(Y_t) K_t^i \, dt \right]$$

$$+ \sum_{i,j=1}^m \left(\frac{1}{2} \frac{\partial^2 f}{\partial y_i \partial y_j}(Y_t) \sum_{k=1}^d H_t^{ik} H_t^{jk} \right) dt \tag{24}$$

where the last term on the right may be unexpected. By contrast (23) does imply

$$dZ_t = \sum_{i=1}^m \left[\sum_{k=1}^d \frac{\partial f}{\partial y_i}(Y_t) H_t^{ik} \circ dW_t^k + \frac{\partial f}{\partial y_i}(Y_t) K_t^i \, dt \right]. \tag{25}$$

If $u: R^d \to R^1$ and its first two partial derivatives are bounded, then $\int_0^t u(W_t) \circ dW_t$ is well defined. This Stratonovitch integral enjoys a very convenient *approximation property*. Namely, let (B_t^n) be the process obtained by interpolating (W_t) linearly between the dyadic points $t = k2^{-n}$, $k = 0, 1, \ldots$. Then (B_t^n) is of bounded variation over finite parameter intervals and $\int_0^t u(B_s^n) \, dB_s^n$ has a natural interpretation as a Stieltjes integral. As $n \to \infty$ this integral converges to $\int_0^t u(W_t) \circ dW_t$, the convergence holding in the sense of $L_1(P)$. For a proof see [19].

(g) Stochastic Differential Equations. The notation of the previous subsection is still in use. A stochastic differential equation of Ito type has the form

$$dX_t = \sigma(X_t, t) \, dW_t + b(X_t, t) \, dt, \qquad X_{t_0} = Z. \qquad (26)$$

Here W is an R^d-valued Wiener process (started at the origin), $\sigma(x, t)$ is an $m \times d$-matrix, $b(x, t) \in R^m$, and Z is an $\mathscr{F}_{t_0}$-measurable random variable in R^m. A *solution* is an R^m valued process (X_t) such that the following three conditions hold:

$$X_t = Z + \int_{t_0}^t \sigma(X_s, s) \, dW_s + \int_{t_0}^t b(X_s, s) \, ds, \qquad t \geq t_0;$$

X_t is $\mathscr{F}_t$-measurable, $t \geq t_0$;

(X_t) has continuous sample functions, with probability one.

Note that the last condition means that with probability one $t \to X_t$ is continuous for all t, $0 \leq t < \infty$.

Some properties of stochastic differential equations will now be given. For proofs, see the references given in the previous subsection.

It was shown by Ito that if the coefficients $\sigma(x, t)$ and $b(x, t)$ are measurable functions satisfying a Lipschitz condition in x, uniformly in x and t, and if they are bounded in absolute value by a constant multiple of $(1 + |x|)$, then (26) has a unique solution. In fact (26) looks like an ordinary differential equation (even more so if one writes formally $dW_t = \dot{W}_t dt$) and the classical Picard argument for existence and uniqueness can be adapted. The solution of

(26) is naturally denoted by $X^{(Z,t_0)}$, or $X^{(x,t_0)}$ if $Z = x$ with probability one. The uniqueness of the solution implies that ($X_t^{(x,s)}$, $x \in R^d$, $s \geqslant 0$) is a Markov process:

$$P\left[X_{t+u}^{(x,s)} \in A | \mathcal{F}_t\right] = P\left[X_u^{(\alpha,t)} \in A\right], \qquad t \geqslant s, u > 0, \qquad \alpha = X_t^{(x,s)}.$$

(27)

The transition probabilities are given by

$$p(x, s; A, t) = P\left[X_t^{(x,s)} \in A\right], \qquad t \geqslant s.$$

When σ and b do not depend on t the transition probabilities will be stationary, $p(x, s; A, t) = p(t - s, x, A)$. In this case, using the Ito formula (24) for a smooth function f gives

$$f(X_t) - f(X_0) = \sum_{i=1}^{m} \sum_{k=1}^{d} \int_0^t \frac{\partial f}{\partial x_i}(X_s) \sigma^{ik}(X_s) \, dW_s^k$$

$$+ \int_0^t \left[\frac{1}{2} \sum_{i,j=1}^{m} a_{ij}(X_u) \frac{\partial^2 f}{\partial x_i \partial x_j}(X_u) + \sum_{i=1}^{m} b_i(X_u) \frac{\partial f}{\partial x_i}(X_u)\right] du$$

where

$$a_{ij}(x) = \sum_{k=1}^{d} \sigma_{ik}(x) \sigma_{jk}(x), \qquad 1 \leqslant i \leqslant m, \qquad 1 \leqslant j \leqslant m.$$

On taking expectations the integrals involving dW_s^k vanish, since they are martingales, and one deduces that the differential generator of the Markov process is

$$L = \frac{1}{2} \sum_{i,j=1}^{m} a_{ij}(x) \frac{\partial^2}{\partial x_i \partial x_j} + \sum_{i=1}^{m} b_i(x) \frac{\partial}{\partial x_i}.$$

It should be observed that the approach to Markov processes via stochastic differential equations leads naturally to a notation in which one has one underlying probability measure P, but a family of processes $X_t^{(x)}$ (or $X_t^{(x,t_0)}$) where the superscript x denotes the position at time zero (or t_0); see Section 0(B). When σ and b are allowed to depend on time, one obtains analogously that the corresponding space time process $\tilde{X}$ has the generator

$$\frac{\partial}{\partial t} + L = \frac{\partial}{\partial t} + \frac{1}{2} \sum_{i,j=1}^{m} a_{ij}(x,t) \frac{\partial^2}{\partial x_i \partial x_j} + \sum_{i=1}^{m} b_i(x,t) \frac{\partial}{\partial x_t}. \quad (28)$$

It is of interest to consider in place of (26) the Stratonovitch stochastic differential equation

$$dX_t = \sigma(X_t, t) \circ dW_t + b(X_t, t)\, dt, \qquad X_{t_0} = Z. \quad (29)$$

Assuming σ smooth, (29) can be shown to be equivalent to the Ito stochastic differential equation

$$dX_t = \sigma(X_t, t)\, dW_t + \left(b(X_t, t) + \tfrac{1}{2}\hat{\sigma}(X_t, t)\right) dt \quad (30)$$

where

$$\hat{\sigma}_i = \sum_{j=1}^{m} \sum_{k=1}^{d} \left(\frac{\partial}{\partial x_j} \sigma_{ik}\right) \sigma_{jk}, \qquad i = 1, 2, \ldots, m,$$

and so the corresponding generator will be

$$L + \sum_{i=1}^{m} \frac{1}{2} \hat{\sigma}_i(x,t) \frac{\partial}{\partial x_i}.$$

For the system (29) one has the Wong-Zakai *approximation property*, analogous to the approximation property for Stratonovitch stochastic integrals. Namely, with B_t^n approximating W_t as before, the solutions of

$$dX_t^{(n)} = \sigma\left(X_t^{(n)}, t\right) dB_t^n + b\left(X_t^{(n)}, t\right) dt, \qquad X_{t_0} = Z_0 \quad (31)$$

converge to the solution of (29). Once again, since B_t^n is differentiable, except at denumerably many points, (31) really can be interpreted as an ordinary differential equation, $dB_t^{(n)} = \dot{B}_t^{(n)} dt$. The proof is again in [19].

Consider now two solutions $(X_t^{(x,t_0)})$ and $(X_t^{(y,t_0)})$ to (32), with the initial conditions

$$X_{t_0}^{(x,t_0)} = x, \qquad X_{t_0}^{(y,t_0)} = y.$$

Let

$$T = \inf\{t: t \geqslant t_0, X_t^{(x,t_0)} = X_t^{(y,t_0)}\} \qquad (32)$$

with $T = \infty$ if not otherwise defined. Then T is a stopping time and, using a version of the uniqueness theorem which allows stopping times for initial times, one deduces that on the set $T < \infty$,

$$P\left[X_{T+u}^{(x,t_0)} = X_{T+u}^{(y,t_0)}, \qquad u \geqslant 0\right] = 1. \qquad (33)$$

This fact will be exploited later.

(h)   Radon-Nikodym Derivatives.   Again let $W_t, \mathscr{F}_t, P$ be as in (f), with W_t in R^d. Let (H_t) be a $(1 \times d)$-matrix valued process satisfying the integrability conditions (a)–(c) given in (f). It is not hard to show that the real valued process

$$R_t = \exp\left\{\int_0^t H_s \, dW_s - \frac{1}{2}\int_0^t |H_s|^2 \, ds\right\} \qquad (34)$$

is a supermartingale. In fact it tends to be a martingale: a sufficient condition for $(R_t, \mathscr{F}_t, 0 \leqslant t \leqslant t')$ to be a martingale is

$$E\left[\exp\left\{(1+\varepsilon)\int_0^{t'} |H_s|^2 \, ds\right\}\right] < \infty \quad \text{for some } \varepsilon > 0. \qquad (35)$$

Indeed $(R_t, \mathscr{F}_t, 0 \leqslant t \leqslant t')$ is a martingale if and only if $ER_{t'} = 1$. In that case one can define a new probability measure by

$$P^*(\Lambda) = \int_\Lambda R_{t'} \, dP, \qquad \Lambda \in \mathscr{F}_{t'}.$$

The remarkable theorem of Cameron-Martin-Girsanov implies that with

$$V_t = W_t - \int_0^t H_s \, ds \qquad (36)$$

one has that $(V_t, \mathscr{F}_t, 0 \leqslant t \leqslant t', P^*)$ is a martingale and Wiener process. That is the V process, which from the viewpoint of the original P is a Wiener process plus a translation, is seen by P^* as a Wiener process!

Apply this to the following pair of Ito equations:

$$dX_t^i = \sigma(X_t^i, t) \, dW_t + b^i(X_t^i, t) \, dt, \qquad i = 1, 2. \qquad (37)$$

Here σ is a $d \times d$ matrix, $X^1, X^2, W, b^1, b^2 \in R^d$. Assume the conditions for existence and uniqueness given in the preceding subsection are satisfied, and furthermore assume that b^1 and b^2 are uniformly bounded and σ^{-1} exists and is uniformly bounded. Set

$$h(x, t) = \sigma^{-1}(x, t)\big[b^2(x, t) - b^1(x, t)\big].$$

Now let $H_s = h(X_s^1, s)$. Then (36) and (37) imply

$$dX_t^1 = \sigma(X_t^1, t) \, dV_t + b^2(X^1, t) \, dt.$$

Thus P^* looking at X^1 sees a process with the same generator as P sees looking at X^2. So if ϕ is any bounded, real valued measurable functional defined on the space of continuous functions from $[0, t']$ into R^d, then

$$\int \phi(X^2) \, dP = \int \phi(X^1) \, dP^* = \int \phi(X^1) R_{t'}, dP. \qquad (38)$$

For proofs refer to Friedman [13].

(i) Zero-one Laws. For certain classes of events one knows a priori that for any Λ in the class, $P(\Lambda)$ must equal either zero or one. Thus in (a) the Blumenthal zero-one law was already encountered. A more elementary example comes from the Borel-Cantelli lemma. There one deals with a sequence of events $\Lambda_1, \Lambda_2, \ldots,$ and the event $\Lambda = \limsup_n \Lambda_n$ (in words: Λ is the event that infinitely many Λ_n occur). The Borel-Cantelli lemma asserts that

$$\sum_{n=1}^{\infty} P[\Lambda_n] < \infty \tag{39}$$

implies

$$P[\Lambda] = 0, \tag{40}$$

and if furthermore the (Λ_i) are independent, then

$$\sum_{n=1}^{\infty} P[\Lambda_n] = \infty \tag{41}$$

implies

$$P[\Lambda] = 1. \tag{42}$$

Of course if the Λ_i are not independent the last implication may fail. Various extensions of the Borel-Cantelli lemma have been found to deal with the case where independence fails. A rather general form was given by Kochen and Stone. Namely, on setting

$$I_{A_i} = 1 \text{ on } A_i, \qquad I_{A_i} = 0 \text{ on } A_i^c$$

and

$$N_k = \sum_{i=1}^{k} I_{A_i},$$

one has that (41) implies

$$P[\Lambda] \geqslant \limsup_{n \to \infty} \frac{(EN_n)^2}{EN_n^2}. \tag{43}$$

This is particularly useful when, on the basis of some zero-one law, one knows $P[\Lambda]$ to equal zero or one, for then the right side of (43) must only be shown to be positive to conclude $P[\Lambda] = 1$. Note that the right side will be positive whenever there exists a constant c such that

$$P\left[\Lambda_i \Lambda_j\right] \leqslant cP\left[\Lambda_i\right]P\left[\Lambda_j\right], \text{ for all } i \text{ and } j.$$

A proof of the Kochen-Stone condition can be found in Spitzer [35], Section 26.

2. HARMONIC FUNCTIONS

Canonical diffusion operators are considered and the harmonic functions corresponding to them. Subsection (A) relates the asymptotic behavior of the characteristic process to the behavior of functions harmonic in some neighborhood of infinity. Some of the problems considered include uniqueness for the exterior Dirichlet problem, multiplicity of the solutions in the nonunique case, Liouville property. In Subsection (B) the classical Dirichlet problem is considered, with special reference to the question of regular and irregular boundary points. Both elliptic and parabolic problems are considered. These questions are closely related to regularity properties of the sample functions of the characteristic process, and some highlights from that theory are presented.

(A) Harmonic Functions Near ∞. For a given differential operator L it is natural to enquire what the class of all harmonic functions defined in all of space looks like, or to ask the same question for important subclasses, such as the space of all positive harmonic functions or all bounded harmonic functions. It turns out that such questions are tied intimately to the behavior of the characteristic diffusion.[2] When the space is R^d, these questions turn into questions about the behavior of harmonic functions near ∞.

---

[2] For details on many of the results given here and related topics the reader may consult R. N. Bhattacharya, *Ann. Probab.*, **6** (1978), 541-553.

Let L be a canonical diffusion operator, with characteristic process (X_t). Consider the Dirichlet problem for an unbounded region D. Assume D open, with a complement D^c which is compact, with nonvoid interior, and suppose D has a smooth boundary ∂D. Let $T = T_{\partial D}$. If $P^x[T < \infty] = 1$ for all x, one can proceed as in (c) of Section 1. Then

$$u(x) = E^x[\phi(X_T)]$$

will certainly be a bounded solution for the Dirichlet problem. To see *uniqueness* consider any other bounded solution $w(x)$. Let r be so big that $D^c \subseteq B_r = \{x: x < r\}$. Then w solves a Dirichlet problem on the bounded region $D \cap B_r$. Letting $T_r = T_{\partial B_r}$, one has

$$w(x) = E^x[\phi(X_T), T < T_r] + E^x[w(X_{T_r}), T_r < T]$$

and by assumption the second term tends to zero as $r \to \infty$.

The process X is *recurrent* if for every open set B, $P^x[T_B < \infty] \equiv 1$. Under our assumptions it is easy to see that if this condition holds for some open B it holds for every open B. So the assumption $P^x[T < \infty] \equiv 1$ is equivalent to recurrence. (That is why D^c was required to have a nonvoid interior.) By contrast X is *transient* if $P^x[\lim_{t \to \infty} |X_t| = \infty] \equiv 1$. In fact X must be either recurrent or transient. To see this let Λ be the event $[\lim_{t \to \infty} |X_t| = \infty]$. It is easily seen, by using the criterion for harmonic functions given in (c) of Section 1, that $h(x) = P^x[\Lambda]$ is harmonic. If $h \equiv 0$, X is recurrent, and if $h \equiv 1$ it is transient. By the strong maximum principle, we need only show that h assumes the value 1 (it must then be identically one) or 0 (it must then be identically zero). Now $h(X_t)$ is a bounded martingale converging to $\chi_\Lambda P^x$-a.s., $h(x) = E^x[\chi_\Lambda]$; see (e) of Section 1. So if $h(x) < 1$, $P^x[h(X_t) \to 0, \Lambda^c] > 0$ and since h is continuous this implies the existence of a point y with $h(y) = 0$.

Obviously it is of interest to know whether a process is recurrent or transient. Such criteria will be discussed later.

In the transient case, for any real number α,

$$u(x) = E^x[\phi(X_T), \quad T < \infty] + \alpha P^x[T = \infty] \tag{1}$$

will provide a bounded solution to the Dirichlet problem. Introducing

$$v_0(x) = P^x[T < \infty],$$

it is clear that $u(x)$ tends to α as $x \to \infty$ if and only if

$$\lim_{x \to \infty} v_0(x) = 0. \tag{2}$$

If for every positive continuous ϕ on ∂D, there exists a bounded solution u of the Dirichlet problem satisfying $u(x) \to 0$ as $x \to \infty$, say that ∞ is *regular*. Evidently this will be the case if (2) holds. A function v is called a *barrier* at ∞ if it is positive and superharmonic in some neighborhood of ∞, and $v(x) \to 0$ as $x \to \infty$. So v_0 is a barrier if and only if (2) holds. It is easy to see that if any barrier exists, (2) must hold. Hence, ∞ *is regular if and only if a barrier at ∞ exists.*

Regularity of ∞ implies transience of X, but not conversely. (Consider for example a diffusion on the line, with $P^x[X_t \to +\infty]$ $= 1$ for all x.) If X is transient, $h(x) = v_0(x)$ will be a positive superharmonic function defined on a neighborhood of ∞ and satisfying $\liminf_{x \to \infty} h(x) = 0$. Such a function will be called a *semibarrier* at ∞. Once again, *a semibarrier at ∞ exists if and only if v_0 is a semibarrier, and this condition is equivalent to the transience of X.* Of course then recurrence is equivalent to not having a semibarrier at ∞, but a more useful condition will now be formulated. A positive superharmonic function defined in a neighborhood of ∞ is an *anti-barrier* if *it tends to infinity as $x \to \infty$. Let us show now that the existence of an antibarrier is necessary and sufficient for recurrence.* The sufficiency is almost immediate: let v be an antibarrier defined in a spherical shell $r_0 \leqslant |x| \leqslant r_1$. Let U be the first time X leaves this shell, x a fixed point inside the shell. Then

$$v(x) \geqslant E^x[v(X_U)];$$

this follows from the fact that $v(X_t)$ is a supermartingale. Let $m(r_1)$ be the infimum of v on $\{X: |x| = r_1\}$. Then

$$v(x) \geqslant m(r_1) P^x[X_U = r_1].$$

Let $r_1 \to \infty$. By assumption $m(r_1) \to \infty$, so $P^x[X_U = r_1] \to 0$, showing X is recurrent. For the converse let

$$T = \inf\{t > 0: |X_t| \le 1\}$$

and choose $g(x) = g(|x|)$ a positive continuous function of $|x|$, increasing to infinity with $|x|$, but sufficiently slowly so that

$$M = \max_{0 \le t \le T} g(X_t)$$

satisfies $E^x M < \infty$ for all x. Let $v(x) = E^x M$. It is immediate that v has the supermean value property on $|x| > 1$ (see (c) of Section 1). Also, using the strong Feller property (see (b) of Section 1) one verifies that v is continuous on $|x| > 1$. So v is superharmonic on $|x| > 1$. Clearly $v(x)$ tends to infinity on $x \to \infty$. So v is the desired antibarrier.

Barriers and antibarriers at ∞ were used by Meyers and Serrin [28] in their study of the exterior Dirichlet problem, and they gave useful sufficient conditions for existence. Introduce the quantities

$$A(x) = |x|^{-2} \sum_{i,j} a_{ij}(x) x_i x_j, \qquad B(x) = \sum_i a_{ii}(x),$$

$$C(x) = 2\sum_i x_i b_i(x), \qquad S(x) = [B(x) - A(x) + C(x)]/A(x).$$

If there exists a continuous function $\varepsilon_1(r)$ on $[r_0, \infty)$ such that $S(x) \ge 1 + \varepsilon_1(|x|)$ for all x such that $|x| \ge r_0$ and

$$\int_{r_0}^\infty t^{-1} \exp\left\{ -\int_{r_0}^t \varepsilon_1(s) s^{-1} ds \right\} dt < \infty,$$

a barrier exists. Explicitly such a barrier is given by $v(x) = F_1(|x|)$ where

$$F_1(r) = \int_r^\infty e^{-I_1(s)} ds, \qquad I_1(s) = \int_{r_0}^s (1 + \varepsilon_1(u)) u^{-1} du.$$

On the other hand, *if there exists a continuous function* $\varepsilon_2(r)$ *on* $[r_0, \infty)$ *such that* $S(x) \leqslant 1 + \varepsilon_2(|x|)$ *and*

$$\int_{r_0}^{\infty} t^{-1} \exp\left\{ -\int_{r_0}^{t} \varepsilon_2(s) s^{-1} \, ds \right\} \, dt = \infty,$$

then an anti-barrier exists. Explicitly $v(x) = F_2(|x|)$ where

$$F_2(r) = \int_{r_0}^{r} e^{-I_2(s)} \, ds, \qquad I_2(s) = \int_{r_0}^{s} (1 + \varepsilon_2(u)) u^{-1} \, du.$$

For the class of operators such that $S(x) = S(|x|)$ the Meyers-Serrin tests are evidently necessary and sufficient. This class contains as a proper subclass those L whose coefficients depend only on $|x|$. Note in particular that for Brownian motion, that is for $L = \frac{1}{2}\Delta$, one has a barrier at ∞ in dimensions 1 and 2, an antibarrier in dimensions greater or equal to three. This is an easy proof of the fact that the Brownian motion process is recurrent if and only if the dimension is at most two. As we have already seen, the exterior Dirichlet problem will have a unique bounded solution if and only if the process is recurrent: so this also gives the familiar fact that for the Laplacian this problem has a unique solution if and only if the dimension is at most two. A remarkable class of L is examined in [28], namely L with $b \equiv 0$ and

$$a_{ij}(x) = \delta_{ij} + g(|x|)|x|^{-2} x_i x_j, \qquad 1 \leqslant i \leqslant d, \quad 1 \leqslant j \leqslant d,$$

with $g > -1$. The coefficient matrix has eigenvalues 1 and $1 + g$, the former with multiplicity $(d - 1)$. So L is elliptic, and uniformly elliptic if g is bounded away from -1 and $+\infty$, and one finds

$$S(x) = \frac{d + g(|x|)}{1 + g(|x|)} - 1 = S(|x|).$$

In particular choosing g constant, one finds that one has a barrier at ∞ (hence X is transient) if and only if $g < d - 2$; otherwise one has an antibarrier (hence X is recurrent). So, even though there are no drift coefficients and the operators are uniformly elliptic, the recurrence properties can be different from those of Brownian motion in the same number of dimensions!

In the transient case, that is when a semibarrier exists, (1) provides a one-parameter family of bounded solutions to the Dirichlet problem. Does this provide all bounded solutions? If not, how does one obtain all bounded solutions? The answers do not depend on D or ϕ, but rather on the class of all bounded harmonic functions in R^d. Thus the answer to the first question will be affirmative if and only if one has the *Liouville property*: all bounded harmonic functions are constant. The Liouville property has an important interpretation in terms of the characteristic process X. For $s \geq 0$ one defines a shift operator θ_s sending random variables Z measurable with respect to $\mathscr{F}_\infty$ into new random variables of the same kind. If

$$Z = g\left(X_{t_1}, \ldots, X_{t_n}\right), \quad \theta_s Z = g\left(X_{t_1+s}, \ldots, X_{t_n+s}\right),$$

and this definition can be extended to cover the case $Z = g(X_{t_1}, X_{t_2}, \ldots)$ or even $Z = g(X)$, where X is the function $t \to X_t$ and g is a functional. The random variable is *invariant* if $Z = \theta_s Z$ for all $s > 0$. A random variable Z is P-invariant if there exists an invariant random variable Z' such that $P^x[Z = Z'] = 1$ for all x; the last relation asserts that Z and Z' are *P-equivalent*. What is of importance is the P-equivalence classes of invariant random variables. An event Λ is P-invariant if χ_Λ is P-invariant. A typical instance of such a Λ is $\Lambda_A = [X_t \in A$ for all sufficiently large $t]$. Between bounded harmonic functions and P-invariant random variables there is a natural correspondence. For if H is a bounded P-invariant random variable, $h(x) = E^x[H]$ defines a bounded harmonic function. Conversely, if h is a bounded harmonic function, $H_t = h(X_t)$ is a bounded P^x-martingale for each x. So H_t converges with probability one to a limit H, satisfying $h(x) = E^x[H]$, according to the martingale convergence theorem. In particular, *the Liouville property holds if and only if all bounded invariant random variables are constant*. In the recurrent case, if h is a bounded harmonic function and H_t the corresponding martingale, the recurrence prevents convergence of H_t unless $H \equiv$ constant: so

the Liouville property must hold. In the transient case it is difficult
to decide when it holds. In fact the study of invariant random
variables reduces to studying invariant sets of the form Λ_A, $A \subseteq R^d$.
Call A an *exit set* if $P^x[\Lambda_A] > 0$ for some x, and say $A \sim B$ if the
symmetric difference of A and B is not an exit set. The Liouville
property becomes: up to $\sim$-equivalence there is no more than one
exit set. When this fails, the question of describing all bounded
harmonic functions reduces to describing all exit sets. For one-
dimensional diffusion this is very easy: in the transient case one of
the sets $(-\infty,0]$ or $[0,\infty)$ must be an exit set, and both may be:
any of these situations can occur. In higher dimensions the problem
is difficult. Even finding sufficient conditions for the Liouville
property is hard. Does the Liouville property hold if $b \equiv 0$ and L is
uniformly elliptic? In [28] an affirmative answer is given under the
restriction that $a_{ij}(x)$ converge to a constant matrix as $x \to \infty$.
Quite recently however N. V. Krylov and M. V. Safonov have
announced far-reaching results which imply that the affirmative
answer always holds.[3] In fact they obtain a Harnack inequality with
the bound depending on the $a_{ij}(x)$ only through the modulus of
ellipticity.

Although bounded harmonic functions correspond to bounded
P-invariant random variables, it is not the case that this correspondence can be extended to unbounded, or positive harmonic functions. The problem of determining all positive harmonic functions
is attacked analytically by means of Martin boundaries or similar
constructions. Though probabilistic methods can be helpful here,
the relation to the underlying process is more remote. As an
interesting example one may consider Brownian motion in R^d,
$d \geqslant 2$, which has no positive harmonic functions other than constants; on the other hand if a constant drift is introduced, corre-
sponding to the operator $\frac{1}{2}\Delta + b \cdot \nabla$, with $b \equiv$ constant $\neq 0$, one
obtains a one-parameter family of positive harmonic functions,
such that all positive harmonic functions are expressible as convex
combinations of these: the precise nature of the Martin boundary

---

[3]See N. V. Krylov and M. S. Safonov, *Dokl. Akad. Nauk* SSSR, **245** (1979),
253–255; English translation in *Soviet Math. Dokl.*, **20** (1979), 253–255.

in this case (or rather the discrete parameter analog) was worked out by Ney and Spitzer [29].

(B) Criteria For Regularity. Now let D be a bounded region and consider the Dirichlet problem for the canonical diffusion operator L: $Lu = 0$ in D, $u = \phi$ on ∂D, with ϕ continuous. Let $T = T_{\partial D}$. Then

$$u(x) = E^x[\phi(X_T)] \tag{1}$$

will be the unique solution, provided ∂D is smooth. If no smoothness assumptions are made, there could be points $g \in \partial D$ such that

$$u(x) \to \phi(y) \text{ as } x \to y \text{ through } D$$

fails. Boundary points for which this relation holds, for all choices of ϕ, are called *regular* boundary points; the others are called *irregular boundary points*. Even though irregular boundary points can exist, (1) always provides a function which serves as a generalized solution. For instance, for $x \in D$,

$$P^x\left[\lim_{s \uparrow T} u(X_s) = \phi(X_T)\right] = 1.$$

It is easy to see what causes the appearance of irregular points. If $y \in \partial D$, u is defined by (1), and x is near y, then $u(x)$ will be near $\phi(y)$ if and only if the exit distribution $P^x[X_T \in \cdot]$ is concentrated near y. This will be the case if ∂D is smooth near y, for then T will be very small with high probability. Under our assumptions slight additional arguments will give more. Let

$$T' = \inf\{t: t > 0, X_t \notin D\}.$$

Then *y is regular if and only if $P^y[T' = 0] = 1$*. Note that according to the Blumenthal zero-one law ((see a) of Section 1) the probability must be either 1 or 0. At this point we have not a test but an interesting alternative formulation of the problem of regularity. For details see Dynkin [11].

A standard example is enlightening. Let $d \geqslant 2$ and let $D = B \setminus \Gamma$, where B is the open unit ball and Γ a radius of that ball. If the Wiener process is started at some point $x \in \Gamma$, it will return to Γ with probability one during any positive time interval if $d = 2$: hence in this case x is regular. However for $d = 3$ it will not return to Γ instantaneously (in fact never), so that the process will in fact not leave D for some positive amount of time, and x is not regular.

For the parabolic operator $\partial / \partial t + L \equiv \tilde{L}$, the space time process $\tilde{X}$ associated with X is the characteristic process. The formulation and solution of the Dirichlet problem, with D now being a region in $(d + 1)$-dimensions, and the discussion of regularity proceeds analogously. Note that if $D = A \times [a, b]$, where A is an open region in R^d, then all boundary points of the form (x, a), with $x \in A$ are clearly irregular: for if $\tilde{X}$ is started at (x, a) it evidently cannot leave D for some positive amount of time. For more information see Doob [9].

For L the Laplacian, Wiener has given a famous criterion for regularity. Probabilistic methods are well suited for obtaining tests of this kind in very general situations, and they lead to interesting new questions and generalizations. Let L, X be as above. For convenience only, let us assume that X is transient. To test whether a given boundary point x is regular we must see whether or not $P^x[T' = 0] = 1$. Choose pairwise disjoint sets $A_1, A_2, \ldots$, such that

$$[T' = 0] = [X_t \text{ enters } A_n \cap D^c \text{ for infinitely many } n], \qquad P^x\text{-a.s.}$$

(E.g., for Brownian motion in dimension at least three, let S_n be a ball with center x, radius r_n, $r_n \downarrow 0$, and $A_n = S_n \setminus S_{n-1}$.) The problem of regularity now becomes, on letting

$$\Lambda_n = [X_t \in A_n \cap D^c \quad \text{for some } t],$$

the problem of deciding whether $P^x[\limsup_n \Lambda_n]$ equals 1 (regularity) or 0 (irregularity). This is amenable to Borel-Cantelli type arguments; see (i) of Section 1. Let

$$h(y, B) = P^y[X_t \in B \quad \text{for some } t].$$

One may hope to obtain

$$x \text{ is regular if and only if } \sum_n h(x, A_n \cap D^c) = \infty. \qquad (2)$$

Indeed, if the series converges, the easy half of the Borel-Cantelli Lemma will give that x is not regular. If one can manage to choose the A_n so that there is enough independence between the Λ_n to apply one of the extensions of Borel-Cantelli discussed in (i) of Section 1, (2) will be established. Even when (2) does hold, it is desirable to transform it into a more concrete form, and sometimes this can be done.

(a) *Elliptic equations.* Let L be a canonical diffusion operator. Assume a potential kernel $g(x, y)$ exists. Let x_0 be the boundary point whose regularity is being considered. Assume $g(x_0, y)$ becomes infinite as y approaches x_0, but $g(x_0, y)$ remains bounded if y remains away from x_0. It is tempting to let

$$A_n = \{x: 2^n \leqslant g(x_0, x) < 2^{n+1}\}. \tag{3}$$

If one can establish (2) in this case one obtains an attractive test for regularity. Since by (c) of Section 1 $h(x_0, A_n \cap D^c)$ is the integral of $g(x_0, y)$ over $A_n \cap D^c$ with respect to the equilibrium measure of that set, and the capacity of $A_n \cap D^c$ is just the total mass of this equilibrium measure,

$$2^n \mathrm{Cap}(A_n \cap D^c) \leqslant h(x_0, A_n \cap D^c) \leqslant 2^{n+1} \mathrm{Cap}(A_n \cap D^c) \tag{4}$$

and one obtains the Wiener *test*

$$\sum_n 2^n \mathrm{Cap}(A_n \cap D^c) = \infty \text{ if and only if } x_0 \text{ is regular.} \tag{5}$$

The extended Borel-Cantelli Lemma can be applied to give (2), provided for some c and all $m \neq n$,

$$P^{x_0}[\Lambda_n \Lambda_m] \leqslant c P^{x_0}[\Lambda_n] P^{x_0}[\Lambda_m]. \tag{6}$$

Indeed it suffices to establish (6) for $|m - n| \geqslant 2$. For in that case divergence of the series will entail that either

$$P^{x_0}[\limsup \Lambda_{2n}] = 1$$

or

$$P^{x_0}[\limsup \Lambda_{2n+1}] = 1,$$

and that is equivalent to

$$P^{x_0}[\limsup \Lambda_n] = 1.$$

To simplify notation let $A'_k = A_k \cap D^c$, $k = 1, 2, \ldots$. It follows from (4) that to conclude (6) it will suffice to establish

$$P^{x_0}[\Lambda_n \Lambda_m] \leqslant c 2^{m+n} \mathrm{Cap}(A'_n) \mathrm{Cap}(A'_m). \tag{7}$$

Now, just as in obtaining (4), estimate for $m \neq n$

$$P^{x_0}[X_t \text{ enters } A'_n \text{ and then enters } A'_m] \leqslant P^{x_0}[\Lambda_n] \sup_{x \in A'_n} P^x[\Lambda_m]$$

$$\leqslant 2^{n+1} \mathrm{Cap}(A'_n) \sup_{x \in A'_n, y \in A'_m} g(x, y) \mathrm{Cap}(A'_m),$$

and a similar estimate holds for the probability that X_t enters A'_m and then A'_n. So to obtain (7) one only needs

$$g(x, y) \leqslant c' g(x_0, y), \quad x \in A'_m, \quad y \in A'_n,$$

$$n \geqslant 1, \quad m \geqslant 1, \quad |n - m| \geqslant 2 \tag{8}$$

for some fixed positive constant c' (depending on c). Condition (8) will hold for example if there exist positive constants k_1, k_2, β such that

$$\frac{k_1}{|x_1 - x_2|^\beta} \leqslant g(x_1, x_2) \leqslant \frac{k_2}{|x_1 - x_2|^\beta}, \qquad x_1 \in R^d, \quad x_2 \in R^d.$$

This holds for Brownian motion in dimension of at least 3 (but a Wiener test in 2 dimensions is obtainable by slight modifications) and many other canonical diffusion operators. The treatment given here parallels Lamperti [24].

(b) *The heat operator and surfaces of revolution.*   In this subsection the operator $\partial/\partial t + \frac{1}{2}\Delta$ will be considered. Let X be the Wiener process in R^d and $\tilde{X}$ the associated space time process. Let ψ be a continuous increasing function of t, defined for $t \geqslant 0$ and vanishing at the origin, $D_\psi = \{(x, t): |\psi(x)| < t\}$ and D_ψ^c the complementary set. Then the question, *Is the origin a regular point for D_ψ?* is equivalent to, *Is*

$$P^{(0,0)}\left[\tilde{X}_t \in D_\psi^c \quad \text{for arbitrarily small positive } t\right]$$

equal to one ? For the case $\psi = \alpha(t \log|\log t|)^{1/2}$ the famous *law of the iterated logarithm* of Khintchine provides the answer: *For $\alpha > \sqrt{2}$ the origin is not regular; for $\alpha \leqslant \sqrt{2}$ the origin is regular.* For more general ψ the answer was obtained independently by Petrowsky, working analytically, and Kolmogorov, proceeding probabilistically. For $\psi(t) = \sqrt{t}\,\phi(t)$, and assuming $\sup \phi = \infty$, ϕ decreasing, the test

$$\int_0^\varepsilon \phi^d \exp\left\{-\frac{1}{2}\phi^2\right\} t^{-1}\,dt = \infty, \quad \text{for } \varepsilon > 0,$$

is necessary and sufficient for the regularity of the origin. The law of the iterated logarithm can be proved in a few lines using Borel-Cantelli arguments as explained above. The Kolmogorov-Petrowski test is more difficult. It is interesting that it can also be obtained by the same kind of argument: only the difficult half of the Borel-Cantelli argument is now delicate and some work is needed to obtain sufficient asymptotic independence. For details see Chung, Erdős, and Sirao [6], where these methods were refined to obtain further results to which we revert in (d) below.

For $d \geqslant 2$, it also becomes of interest to ask whether the origin is regular for the "exterior" region D_ψ^c. The necessary and sufficient condition for regularity becomes, for $\psi(t) = \sqrt{t}\,\phi(t)$, $\phi(0) = 0$, ϕ increasing

$$\int_0^\varepsilon \phi^{d-2}(t) t^{-1}\,dt = \infty, d \geqslant 3; \qquad \int_0^\varepsilon |\log \phi(t)|^{-1} t^{-1}\,dt = \infty, \quad d = 2$$

as shown by Dvoretsky and Erdős [10], and Spitzer [34]. Thus for $d = 3$, the origin is not regular for the cone $\{(x, t): |x| \leqslant ct\}$.

(c) *The parabolic case.* It is natural to ask whether in the parabolic case there does not exist a kind of Wiener test for regularity which applies in more general situations than those described in the preceding subsection.

In fact the Wiener test (5) makes sense, but even for the heat operator it is an open problem whether or not the criterion is correct in this case.[4] In any case the proof given in the elliptic case does not work, for if the regions A_n are chosen as in (3), the condition (5) does not hold. The potential kernel now is just $p(x, s; y, t)$ as explained at the end of (c) in Section 1. Instead one may try choosing sets A_n in a different manner. Consider this, still for the heat equation in one dimension, though the arguments easily generalize. Let D be an open region in $R^1 \times [0, \infty)$ with the origin as boundary point. Letting X_t be the Wiener process,

$$T = \inf\{t: t > 0, (X_t, t) \notin D\},$$

our problem is equivalent to deciding whether $P^0[T = 0] = 1$ (regularity) or 0 (irregularity). From what is known from (b) above one may assume

$$D \subseteq \{(x, t): |x| < \sqrt{t \log t}\}.$$

Next, choose a decreasing sequence of positive real numbers τ_n and let

$$A_n = \{(x, t): \tau_n \leqslant t < \tau_{n-1}\}.$$

It is easily checked that condition (6) will hold if

$$\tau_n = \tau_{n-1} / \left(2(\log \tau_{n-1})^2\right).$$

---

[4]Meanwhile a positive solution has been obtained by L. C. Evans and R. F. Gariepy.

Hence *the condition* (2) *will provide a test* for regularity of the origin. An analytic proof of such a result was sketched by Landis [25].

(d) *Sample functions.* The results of (b) give the modulus of continuity at the origin for the Wiener process W_t in R^d. The law of the iterated logarithm says that $\alpha(t\log|\log t|)^{1/2}$ serves as a modulus of continuity if and only if $\alpha > \sqrt{2}$; and the Kolmogorov-Petrowski criterion gives a more refined result. It will be interesting to consider two closely related problems, even though it is not apparent that they have any application to differential equations.

The first problem concerns a *uniform* modulus of continuity for W_t. It was shown by Paul Lévy that $\alpha(t|\log t|)$ provides such a modulus if and only if $\alpha > \sqrt{2}$; in fact,

$$P^0\left[\limsup_{\substack{t_2 - t_1 = \varepsilon \downarrow 0 \\ 0 \leqslant t_1 < t_2 < 1}} \left(W_{t_2} - W_{t_1}\right)\left(2\varepsilon|\log \varepsilon|\right)^{-1/2} = 1\right] = 1.$$

Later Chung, Erdős, and Sirao [6] showed that if ϕ is a positive continuous function on $(0, \infty)$ with $\phi(t)\uparrow 0$ as $t\downarrow 0$ then

$$P^0[\text{there exists a positive } \delta \text{ such that}$$

$$|W_t - W_s| \leqslant \sqrt{t - s}\,\phi(t - s), \quad 0 \leqslant s \leqslant t \leqslant s + \delta \leqslant 1]$$

equals zero or one according as

$$\int_0^\varepsilon t^{-2}\phi^{d+2}(t)e^{-\phi^2(t)/2}\,dt$$

diverges or converges. The method of proof is a very refined version of the Borel-Cantelli techniques explained above. For a fixed $t_0 > 0$, it is natural to enquire about the *two sided local modulus of continuity*, that is, in considering $(W_{t_0+h} - W_{t_0})$ h can be a small positive or negative increment. Jain and Taylor [21] showed that for this problem the integral test of Chung, Erdős, Sirao must be modified by replacing t^{-2} by t^{-1}. This test is for fixed t_0, and since it is a probability one result, one can apply Fubini's theorem to deduce that if the integral converges, $\sqrt{h}\,\phi(h)$ will provide an a.s. local modulus continuity at each fixed t outside a random exceptional set of Lebesgue measure zero. Indeed, if ϕ is such that the Chung-Erdős-Sirao integral diverges so that $\sqrt{h}\,\phi(h)$ is not a uniform modulus, it is easy to see that the exceptional set is not void. How big is it? Under suitable regularity conditions on ϕ, Kôno [22] has proved the following beautiful result: *For any Hausdorff measure function $h(t)$, the exceptional set will have h-Hausdorff measure equal to zero or infinity with probability one depending on whether*

$$\int_0^\varepsilon t^{-2}\phi^{d+2}(t)\,e^{-\phi^2(t)/2}h(t)\,dt$$

converges or diverges.

3. GEOMETRY OF PATHS

The results of this section all concern solutions of problems involving diffusion operators in which the characteristic processes are used, and particularly the geometry of the sample paths X_t is exploited.

The first application, in Subsection (A), explains an idea of Conway [7] for proving a result about solutions to the parabolic Dirichlet problem by using stochastic integral equations in a nice way. The application particularly recommends itself because of its great simplicity.

Subsection (B) is devoted to an exposition of the Stroock-Varadhan support theorem [36] for operators L of the form (1.1). The operator is not assumed to be positive definite, and, as explained, this result leads to a strong maximum principle for such operators. It is interesting to note that earlier "purely analytic" approaches to the strong maximum principle already had a probabilistic flavor.

Finally, Subsection (C) gives a brief explanation of the work of Ventcel and Freidlin [39] on dynamical systems with small random perturbations and some of its striking applications to singular perturbation problems. Friedman's monograph [13] also contains an exposition of this work.

(A) Total Variation of Solutions. With x a real variable, consider the equation

$$\frac{\partial u}{\partial t}(x,t) + b(x,t)\frac{\partial u}{\partial x}(x,t) + \frac{1}{2}\sigma^2(x,t)\frac{\partial^2 u}{\partial x^2}(x,t) = 0. \quad (1)$$

In [7] Conway gave an application of the stochastic calculus to study the variation of $u(x,s)$ as a function of x for fixed s. Actually in [7] equations with more general right-hand sides are considered, but the idea can be illustrated for the above equation. For more complete results and discussion of their significance see the original paper.

Suppose then that b and σ satisfy conditions for the existence and uniqueness of the Ito stochastic differential equation

$$dX_t = \sigma(X_t, t)\,dW_t + b(X_t, t)\,dt, \qquad X_s = x \quad (2)$$

as discussed in (g) of Section 1.

Suppose u satisfies (1) inside an open set in $R^1 \times (-\infty, \infty)$ and that u is continuous on the closure of U. Assume the boundary of U has the form $L \cup J$, where the lower boundary is a horizontal line segment L, contained in $R_1 \times \{s\}$, and the rest of the boundary J is a Jordan arc homeomorphic to $[0,1]$, so that

$$T^{(x,s)} = \inf\{t : X^{(x,s)} \in J\}$$

satisfies

$$P[T^{(x,s)} < \infty] = 1 \quad \text{for all } (x,s) \in L.$$

Here of course $X^{(x,s)}$ is the solution to (2). Then the following result holds:

Total variation of u over $L \leqslant$ total variation of u over J.   (3)

The proof follows from the fact that if, as usual, $X_t^{(x,s)}$ denotes the solution of (2), *then $x < y$ implies $X_t^{(x,s)} \leqslant X_t^{(y,t)}$ for all $t \geqslant s$, with probability one.* This was shown in (g) of Section 1; see (1.33). Now use the representation

$$u(x, s) = E\left[u\left(X_{T^{(x,s)}}^{(x,s)}, T^{(x,s)}\right)\right].$$

If $x_1 < x_2 < \cdots < x_n$ with each $(x_i, s) \in L$, the points $X_{T^{(x_i,s)}}^{(x_i,s)}, T^{(x_i,s)})$ will be in ascending order on J. Then

$$\sum_{i=1}^{n} |u(x_i, s) - u(x_{i-1}, s)|$$

$$\leqslant E\left[\sum_{i=1}^{n} \left|u\left(X_{T^{(x_i,s)}}^{(x_i,s)}T^{(x_i,s)}\right) - u\left(X_{T^{(x_{i-1},s)}}^{(x_{i-1},s)}, T^{(x_{i-1},s)}\right)\right|\right]$$

and the right side is bounded above by the total variation of u over J. Hence (3) follows.

(B) Support Theorem and Maximum Principle. Let

$$L = \frac{1}{2} \sum_{i=1}^{d} a_{ij}(x, t) \frac{\partial^2}{\partial x_i \, \partial x_j} + \sum_{i=1}^{d} b_i(x, t) \frac{\partial}{\partial x_i}. \tag{1}$$

A strong maximum principle asserts that for every open set $G \subseteq R^d \times [0, \infty)$ and every $(x, t) \in G$ there exists a subset $G_{x,t}$ of G with the property that $(\partial / \partial t + L)f \geqslant 0$ on G and $f(x, t) = \sup\{f(y, s) : (y, s) \in G\}$ imply $f \equiv f(x, t)$ on $G_{x,t}$. Evidently any subset of $G_{x,t}$ has the same property: a useful strong maximum principle consists of an explicit description of a big, preferably maximal $G_{x,t}$ with the desired property. Assuming bounded continuous coefficients in L and that (a_{ij}) is everywhere positive definite, the Nirenberg strong maximum principle asserts that $G_{x,t}$ can be taken to be the closure in G of the class of points (y, s) such that $s \geqslant t$ and there exists a continuous map ϕ from $[t, s]$ into R^d such that

$$\phi(t) = x, \phi(s) = y, \quad \text{and} \quad (\phi(u), u) \in G \quad \text{for all } u \in (t, s).$$

The maximum principle which follows from the support theorem of

Stroock and Varadhan [36] includes this as a special case, with the additional information that the indicated choice of $G_{x,t}$ is maximal. The main virtue of the support-theorem approach is that it provides a satisfactory maximum principle in case (a_{ij}) is not required to be everywhere positive definite.

The connection between a maximal principle for L and a support theorem for the associated diffusion is not hard to see. Let X be the characteristic process for L, $\tilde{X}$ the corresponding space-time process, and $P^{(x,s)}$ the probability measure corresponding to $\tilde{X}$ started at (x, s). Define $\text{supp}[P^{(x,s)}]$ as the set of all continuous maps ϕ from $[s, \infty)$ into R^d which satisfy

$$P^{(x,s)}\left[\sup_{s \leqslant u \leqslant t} |X_u - \phi_t| < \varepsilon \right] > 0$$

for every finite $t > s$, $\varepsilon > 0$. In case f satisfies $\left(\dfrac{\partial}{\partial t} + L \right)f \geqslant 0$ on G, where the relevant derivatives of f are assumed to exist and be bounded and continuous on G, $-f(\tilde{X}_{t \wedge T})$ will be a supermartingale. So for $T = \inf\{t \colon \tilde{X}_t \notin G\}$

$$f(x, s) \leqslant E^{(x,s)}\left[f(\tilde{X}_{t \wedge T}) \right], \qquad t \geqslant s.$$

Hence if $f(x, s) = \sup\{f(y, t) \colon (y, t) \in G\}$, it is clear that one may take $G_{x,s}$ to be the class of (y, t) such that $t \geqslant s$ and there exists a $\phi \in \text{supp}[P^{(x,s)}]$ such that the graph of ϕ is contained in G, $\phi(0) = x$, $\phi(t) = y$. Nor is it hard to see that this choice gives a maximal $G_{x,s}$. So obtaining a sharp maximum principle is equivalent to describing $\text{supp}[P^{(x,s)}]$.

Suppose that X_t is a diffusion process corresponding to the stochastic differential equations of Stratonovitch type

$$dX_t = \sigma(X_t, t) \circ dW_t + b(X_t, t)\, dt, \quad X_s = x, \quad s \geqslant 0, \quad x \in R^d.$$

$$(2)$$

Here the notation is as in (1.29). Now however take $m = d$ (so that σ is a $d \times d$ matrix). Further, assume now that b and the first spatial derivative of σ are uniformly Lipschitz continuous in x.

Then (see (g) of Section 1) X corresponds to the operator given by

$$Lu = \frac{1}{2} \sum_{i,j,l=1}^{d} \sigma^{il}(x,t) \frac{\partial}{\partial x_i} \left[\sigma^{jl}(x,t) \frac{\partial}{\partial x_j} u \right] + \sum_{i=1}^{d} b_i(x,t) \frac{\partial u}{\partial x_i}.$$

(3)

The support problem for this case is solved in [36]. However, given an operator in the form (1), it is not easy to tell when it can be put in the form (3), with the imposed regularity conditions. For this reason Stroock and Varadhan extended their work in [37] to cover operators in the form (1). This extension will not be discussed here.

Consider then the L of (1). Let $\Phi_{x,s}$ be the class of continuous functions ϕ from $[s, \infty)$ into R^d for which there exists a piecewise constant ψ such that

$$\phi(t) = x + \int_s^t \sigma(\phi(u), u) \psi(u)\, du + \int_s^t b(\phi(u), u)\, du, \qquad t \leqslant s.$$

(4)

The support theorem asserts that

$$\mathrm{supp}[\, P^{(x,s)}\,] = \overline{\Phi}_{x,s},$$

(5)

where the bar denotes topological closure.

The inclusion $\mathrm{supp}[P^{(x,s)}] \subseteq \overline{\Phi}_{x,s}$ follows by approximating the equation (2) as was done in (1.31). The approximations are evidently in $\overline{\Phi}_{x,s}$, and then one applies the convergence result mentioned in connection with (1.31).

The inclusion $\overline{\Phi}_{x,s} \subseteq \mathrm{supp}[P^{(x,s)}]$ is a consequence of the following fact proved in [36]: If ϕ satisfies (4) and ψ is the derivative of a C^2 function Ψ from $[s, \infty)$ to R^d with $\Psi_s = 0$, then for any finite T, $\varepsilon > 0$,

$$\lim_{\delta \downarrow 0} P\left[\sup_{s \leqslant t \leqslant T} |X_t - \phi_t| < \varepsilon \,\middle|\, \sup_{s \leqslant t \leqslant T} |W_t - \Psi_t| < \delta \right] = 1.$$

(6)

In fact this assertion can be deduced without much difficulty from the special case obtained by setting $\Psi \equiv 0$. This special case is of intrinsic interest. The proof for this case takes a certain amount of careful estimating, using the stochastic calculus. Details are given in [36]. Since the conditioning event in (6) is easily shown to have positive probability, $\phi \in \text{supp}[P^{(x,s)}]$ follows.

(C) Small Perturbations. For $\varepsilon \geqslant 0$ let

$$L_\varepsilon = \varepsilon^2 \frac{1}{2} \sum_{i,j=1}^d a_{ij}(x) \frac{\partial^2}{\partial x_i \partial x_j} + \sum_{i=1}^d b_i(x) \frac{\partial}{\partial x_i}, \qquad x \in R^d.$$

A class of questions which has been treated extensively by analytic methods concerns the behavior of solutions of problems involving L_ε as $\varepsilon \downarrow 0$; in particular there arises the question of convergence of solutions to solutions of related problems involving L_0. It turns out that probabilistic methods are very fruitful in this context. Some introduction to the ideas of Ventcel and Freidlin [39] will be presented. Independently closely related results were developed by Varadhan [40]. An exposition and some further results can be found in [13].

It will be assumed that the coefficients a_{ij} and b_i are bounded and uniformly Lipschitz continuous, and that (a_{ij}) is uniformly positive definite. Corresponding to L_ε one has the characteristic process determined by the Ito equation

$$dW_t^{\varepsilon,x} = \varepsilon\sigma(X^{\varepsilon,x})dW_t + b(X^{\varepsilon,x})dt, \qquad X_0^{\varepsilon,x} = x,$$

where $\sigma\sigma^* = a$, σ^* being the transpose of σ, and W is the Wiener process. For $\varepsilon = 0$, this is a deterministic dynamical system, giving the characteristics corresponding to L_0. For fixed $t_0 > 0$, one may expect that for small ε, $X_t^{\varepsilon,x}$ is unlikely to deviate much from $X_t^{0,x}$ for $0 \leqslant t \leqslant t_0$. The basic estimates of Ventcel and Freidlin give good estimates for the probability of having a preassigned deviation from $X_t^{0,x}$. This will be made precise now.

With $t_0 > 0$ fixed, let C be the class of continuous functions ϕ from $[0, t_0]$ into R^d endowed with the uniform topology. For $x \in R^d$, define $\|\cdot\|_x$ by

$$\|y\|_x^2 = y^* a^{-1}(x) y = |\sigma^{-1}(x)y|^2,$$

where y^* is the transpose of y, and $y \in R^d$. Then define, for $\phi \in C$, $I(\phi)$ to equal ∞ if ϕ is not absolutely continuous; otherwise

$$I(\phi) = \int_0^{t_0} \|\dot{\phi}(t) - b(\phi(t))\|_{\phi(t)}^2 \, dt,$$

where $\dot{\phi}$ is the derivative with respect to t. Concentrating now on $0 \le t \le t_0$, each path $X^{(\varepsilon, x)}$, being continuous, will be an element of C, at least a.s. The *basic estimates* can now be stated.

The lower bound. For any open subset G of C

$$\varliminf_{\varepsilon \downarrow 0} 2\varepsilon^2 \log P[X^{(\varepsilon, x)} \in G] \ge -\inf\{I(\phi) : \phi(0) = x, \phi \in G\}.$$

The upper bound. For any closed subset F of C

$$\varlimsup_{\varepsilon \downarrow 0} 2\varepsilon^2 \log P[X^{(\varepsilon, x)} \in F] \le -\inf\{I(\phi) : \phi(0) = x, \phi \in F\}.$$

The upper bound result is established by an ingenious argument of some length. On the other hand the lower bound is obtained quite quickly, and it seems instructive to sketch the argument. To obtain the lower bound it suffices to consider the case where G is an open ball of radius δ centered at $\phi \in C$ and to prove the lower bound inequality in this case with the right side replaced by $I(\phi)$. Let $Y^{(\varepsilon)} = X^{(\varepsilon, x)} - \phi$, so that

$$Y_t^{(\varepsilon)} = \varepsilon \int_0^t \sigma(Y_s^{(\varepsilon)} + \phi(s)) \, dW_s + \int_0^t [b(Y_s^{(\varepsilon)} + \phi(s)) - \dot{\phi}(s)] \, ds.$$

Evidently $X^{(\varepsilon, x)}$ remaining in G up to time t_0 is equivalent to $|Y_t^{(\varepsilon)}| < \delta$ for all $t \in [0, t_0]$. Now let $Z^{(\varepsilon)}$ solve

$$Z_t^{(\varepsilon)} = \varepsilon \int_0^t \sigma(Z_s^{(\varepsilon)} + \phi(s)) \, dW_s, \qquad z_0^{(\varepsilon)} = 0.$$

Now one can apply (1.38) to obtain

$$\int_{\substack{\sup \\ t \le t_0} |Y_t^{(\varepsilon)}| < \delta} dP = \int_{\substack{\sup \\ t \le t_0} |Z_t^{(\varepsilon)}| < \delta} R \, dP \tag{1}$$

where by (1.34)

$$R = \exp\left\{ \frac{1}{\varepsilon} \int_0^{t_0} \left(\sigma^{-1}\left(Z_s^{(\varepsilon)} + \phi(s) \right) \right) \left[b\left(Z_s^{(\varepsilon)} + \phi(s) \right) - \dot{\phi}(s) \right] \right)^* dW_s \right.$$

$$\left. - \frac{1}{2\varepsilon^2} \int_0^{t_0} |\sigma^{-1}\left(Z_s^{(\varepsilon)} + \phi(s) \right) b\left(Z_s^{(\varepsilon)} + \phi(s) \right) - \dot{\phi}(s)|^2 \, ds \right\}.$$

Set

$$\Lambda = \left[\sup_{t \le t_0} |Z_t| < \delta \right].$$

The proof is easily completed by noting that on Λ the second integral in the exponential is close to $I(\phi)$, and as ε tends to zero the first integral becomes unimportant relative to the second one because of the factors ε^{-1} and ε^{-2} respectively. Finally it must be noted that $P[\Lambda] > \frac{1}{2}$ (say) for small ε, so that the logarithm of the right side of (1) can be easily estimated to obtain the desired result.

The basic estimates have been used to treat a variety of problems in partial differential equations. Here only the Dirichlet problem

$$L_\varepsilon u_\varepsilon = 0 \text{ in } D, \qquad u_\varepsilon = \phi \text{ on } \partial D,$$

where D is a connected open set with smooth boundary ∂D and with compact closure, and ϕ is a continuous function on the boundary, will be considered. The question to be considered is the behavior of u^ε as $\varepsilon \downarrow 0$. With $T^{(\varepsilon, x)}$ denoting the first time $X^{(\varepsilon, x)}$ hits ∂D, one has the representation

$$u_\varepsilon(x) = E\left[\phi\left(X_{T^{(\varepsilon, x)}}^{(\varepsilon, x)} \right) \right]$$

so that the question is equivalent to asking about the limiting distribution of the exit point $X_{T^{(\varepsilon, x)}}^{(\varepsilon, x)}$; hence this problem is also referred to as the exit problem. There are various cases that arise. The simplest one is that in which D is convex and the (deterministic!) limiting system $X^{(0, x)}$ is such that, for each $x \in D$, $X^{(0, x)}$ hits the boundary at a finite time $T^{(0, x)}$. It seems intuitively clear that, during the finite time interval $[0, T^{(0, x)}]$, $X_t^{(\varepsilon, x)}$ will remain close to $X_t^{(0, x)}$ with high probability if ε is small. This follows easily from the basic estimates or can be proved by more elementary arguments. So in this case $u^\varepsilon(x) \to \phi(X_{T^{(0, x)}}^{(0, x)})$. Consider now the following case which already embodies most of the essential difficulties. In D there is an equilibrium point 0 for the dynamical system $X^{(0, x)}$ to which all trajectories are attracted, and on ∂D the vector field b points inward. So for $x \in D$, $X_t^{(0, x)}$ approaches 0 as $t \to \infty$, with $X_t^{(0, x)}$ never leaving D. So in this case $X_t^{(0, x)}$ never gets to the boundary. For ε small but positive the paths of $X_t^{(\varepsilon, x)}$ have a tendency to follow $X_t^{(0, x)}$ for a long time. Eventually however, they must reach ∂D. How does this occur? Using the basic estimates, Ventcel and Freidlin were able to answer this question. Introduced above were the space C of continuous functions from $[0, t_0]$ into R^d and an associated functional $I(\phi)$. It is now desirable to make explicit the parameter t_0, and to write C_{t_0} and $I_{t_0}(\phi)$. The following function is called a *pseudo-potential*:

$$V(y) = \inf\{I_{t_0}(\phi): \phi \in C_{t_0}, 0 \leqslant t_0 < \infty, \phi(0) = 0, \phi(t_0) = y\}.$$

In some sense $I_{t_0}(\phi)$ gives the energy it takes to travel along the parametrized curve ϕ, and $V(y)$ is the minimal energy required to get from 0 to y. Using the basic estimates, it is not hard to show that *provided the minimum of $V(y)$ over ∂D is attained at a unique point $y \in \partial D$ the distribution of $X_{T^{(\varepsilon, x)}}^{(\varepsilon, x)}$ concentrates around y_0*, i.e.,

$$\lim_{\varepsilon \downarrow 0} u^\varepsilon(x) = \phi(y_0), \qquad x \in D.$$

This is an intriguing result for which up to now there is no known nonprobabilistic proof. When the proviso that the minimum of V on ∂D is attained at a unique point is violated, it appears hard to say what will happen. There is an interesting class of operators for which this question has been answered. This is the class for which L_ε can be represented as

$$L_\varepsilon = \varepsilon^2 \sum_{j,k=1}^{d} \frac{\partial}{\partial x_j}\left[a_{jk}(x)\frac{\partial}{\partial x_k}\right] + \sum_{j=1}^{d} \tilde{b}_j(x)\frac{\partial}{\partial xj}$$

with $\tilde{b}(x)$ obtained from a function $\psi(x)$ by multiplying the gradient of ψ on the left by the matrix $(a_{ij}(x))$. In that case ψ turns out to be a constant multiple of V, and the limit of $u^\varepsilon(x)$ is shown to exist and is identified in [8], even when there are many points on ∂D where ψ attains its minimum. The arguments in [8] are purely analytic and do not depend on the results of Ventcel and Freidlin. Everything depends on the relation

$$e^{\psi/\varepsilon}L_\varepsilon u = \varepsilon \sum_{j,k=1}^{d} \frac{\partial}{\partial x_k}\left[e^{\psi/\varepsilon}a_{jk}\frac{\partial u}{\partial x_j}\right];$$

for $u = u_\varepsilon$ the left side vanishes, and applying the divergence theorem gives information about an integral over the boundary. For details and reference to earlier work see [8].

4. OTHER OPERATORS

Probabilistic methods can also be fruitfully employed in dealing with differential operators not of the form (1.1). In Subsection (A) an introduction to random evolution is given with applications to some problems in differential equations. There is a sizeable literature revolving around these topics, but since there exist the highly readable surveys by Hersh [16], [17], the reader is urged to consult these for further information.

Probabilistic treatments of problems in nonlinear diffusion are the content of Subsection (B). Again, there is a rapidly growing literature on this subject. Here only some results related to the Kolmogorov-Petrowski-Piscounov equation will be briefly discussed.

(A) Random Evolutions. Random evolutions are sometimes studied because they appear to be good mathematical models for physical situations. Here the emphasis will be entirely on the use of random evolutions to approach some problems in partial differential equations. There exist excellent survey articles on random evolutions by Hersh [16], [17], so this discussion will be brief.

Suppose $Y_t^{(i)}$ is a diffusion in R^d with generator $L^{(i)}$, $i = 1, 2, \ldots, n$. Let X_t be a Markov chain with stationary transition probabilities

$$p_{ij}(t) = P^i[X_t = j], i = 1, 2, \ldots, n, \quad j = 1, 2, \ldots, n.$$

Let $q_{ij} = p'_{ij}(0)$, $Q = (q_{ij})$. (This is the kind of jump process discussed in Section 1(a).) Now construct a new process Y_t as follows: given initial states i for X and y for $Y^{(i)}$, let Y_t agree with Y_t^i started at y up to the first time τ_1 that X jumps. If X jumps to j, Y_t moves on continuously, now however using the transition law of the $Y^{(j)}$ process. This continues up to the time τ_2 of the next jump of the X process, etc. Then $Z_t = (Y_t, X_t)$ is a Markov process, and if f is a real valued, bounded, measurable function on $R^d \times \{1, 2, \ldots, n\}$, with $f(\cdot, i)$ in the domain of $L^{(j)}$ for all i and j, then f is in the domain of the generator L of Z,

$$Lf(y, i) = L^{(i)}f(y, i) + Qf(y, i)$$

where of course $L^{(i)}$ acts on the first argument of f, Q on the second. Hence, as can be seen from (1.4),

$$u(y, i; t) = E^{(y, i)}[f(Z_t)] \tag{1}$$

solves

$$\frac{\partial u}{\partial t}(y, i; t) = L^{(i)}u(y, i; t) + Qu(y, i; t), \qquad t > 0;$$

$$u(y, i; 0) = f(y, i).$$

Specialize now to the case $n = 2$,

$$Q = \begin{pmatrix} -a & a \\ a & -a \end{pmatrix}$$

with Y^1 and Y^2 uniform motion in R^1 with velocity c and $-c$ respectively. Here $a > 0$. Writing $u_i(y, t)$ and $f_i(y)$ in place of $u(y, i; t)$ and $f(y, i; t)$ results in

$$\frac{\partial u_1}{\partial t} = c\frac{\partial u_1}{\partial x} - au_1 + au_2$$

$$\frac{\partial u_2}{\partial t} = -c\frac{\partial u_2}{\partial x} + au_1 - au_2 \qquad (2)$$

with

$$u_1(x,0) = f_1(x), \qquad u_2(x,0) = f_2(x).$$

Now let N_s be the Poisson process with intensity a, started at $N_0 = 0$, (see (a) of Section 1) defined on some probability space $(\Omega, \mathcal{F}, P)$. Set

$$S_t = \int_0^t (-1)^{N_s} ds.$$

Then for $K_t^{(i)} = (i + N_t) \pmod 2$, the random variable $(y + S_t, K_t^{(i)})$ defined on $(\Omega, \mathcal{F}, P)$ has the same distribution as Z_t with respect to the probability measure $P^{(y,i)}$, and the solution (1) can be written

$$u_1(y, t) = E\left[f_1(y + cS_t)\right] P\left[N_t \text{ is even}\right]$$

$$+ E\left[f(y + cS_t)\right] P\left[N_t \text{ is odd}\right], \qquad (3)$$

$$u_2(y, t) = E\left[f_2(y - cS_t)\right] P\left[N_t \text{ is odd}\right]$$

$$+ E\left[f_2(y - cS_t)\right] P\left[N_t \text{ is even}\right]. \qquad (4)$$

Now observe that $v = u_1 + u_2$ satisfies the telegrapher's equation:

$$\frac{\partial^2 v}{\partial t^2} = -2a\frac{\partial v}{\partial t} + c^2\frac{\partial^2 v}{\partial x^2}. \qquad (5)$$

Now the derivation of the Kac *representation* of v in terms of a solution to the wave equation will be obtained. Setting

$$w(x,t) = f_1(x + ct) + f_2(x - ct),$$

certainly w solves

$$\frac{\partial^2 w}{\partial t^2} = c^2 \frac{\partial^2 w}{\partial x^2} \tag{6}$$

and satisfies the initial conditions

$$w(x,0) = v(x,0), \qquad \frac{\partial w}{\partial t}(x,0) = \frac{\partial v}{\partial t}(x,0).$$

Adding (3) and (4) leads to

$$v(y,t) = E\left[w(y, S_t); N_t \text{ is even}\right] + E\left[w(y, -S_t); N_t \text{ is odd}\right]$$

$$= E\left[w(y, S_t)\right] + E\left[w(y, S_t) - w(y, -S_t); N_t \text{ is odd}\right],$$

and since the distribution of S_t on $[N_t$ is odd] is symmetric around zero, the last expectation vanishes. This gives the desired representation

$$v(y,t) = E\left[w(y, S_t)\right]. \tag{7}$$

The derivation of the Kac representation given here follows Heath [15]; see also the discussion in Griego and Hersh [14]. As emphasized by these authors, the argument can be extended without essential change to the d-dimensional case. Even more generally the Laplacian can be replaced by a suitable operator A^2, and the connection between the "abstract telegrapher's equation"

$$\frac{\partial^2 w}{\partial t^2} = A^2 w - 2a \frac{\partial w}{\partial t}$$

and the "abstract wave equation" established, provided A is the generator of a strongly continuous group of bounded linear operators on a Banach space.

Griego and Hersh [14] exploited the representation (7) in an interesting manner. Replace a by a/ε^2, and c by $1/\varepsilon$, with $\varepsilon > 0$. Then (7) becomes,

$$v^{\varepsilon}(y,t) = E\left[w^{\varepsilon}(y, S_t^{\varepsilon})\right] \tag{8}$$

and (5) implies

$$\varepsilon^2 \frac{\partial^2 v^{\varepsilon}}{\partial t^2} = -2a\frac{\partial v^{\varepsilon}}{\partial t} + \frac{\partial^2 v^{\varepsilon}}{\partial x^2},$$

$$v^{\varepsilon}(x,0) = v(x,0), \qquad \frac{\partial v^{\varepsilon}}{\partial t}(x,0) = \frac{\partial v}{\partial t}(x,0), \tag{9}$$

and it is clear that $w^{\varepsilon}(y,t) = w(y, t/\varepsilon)$. Now, using the central limit theorem, it can be shown that the process (S_t^{ε}) converges in distribution as $\varepsilon \downarrow 0$ to $(a^{-1/2}W_t)$, where W is the Wiener process started at the origin. One obtains

$$\lim_{\varepsilon \to 0} v^{\varepsilon}(y,t) = E\left[w\left(y, a^{-1/2}W_t\right)\right]$$

$$= E\left[f_1\left(y + a^{-1/2}W_t\right) + f_2\left(y - a^{-1/2}W_t\right)\right] \equiv v^0(y,t).$$

From the form of $v^0(y,t)$ follows that

$$0 = -2a\frac{\partial v^0}{\partial t} + \frac{\partial^2 v^0}{\partial x^2}; \qquad v^0(x,0) = v(x,0). \tag{10}$$

Again this follows from (1.4). So *the solution of* (9) *converges to that of* (10). As discussed in Hersh [17] and Griego and Hersh [14], this result is classical, but again the present method can be used in quite general settings, leading to new results.

As shown in Hersh and Papanicolaou [18], Hersh [16], formula (1) can usefully be viewed in a more general setting. For this purpose consider a metric space S and a Banach space B. For every $x \in S$, $V(x)$ is to be a densely defined linear operator from B into itself; t is a real parameter. Q is to be the infinitesimal generator of an S-valued Markov process, except that it will be considered as operating on functions $f\colon S \to B$ (rather than real valued functions) as follows: if also $g\colon S \to B$, $Qf = g$ if, in the topology of B,

$$\lim_{t \to 0} \frac{1}{t} E[f(X_t) - f(x)] = g(x).$$

The equation to be considered is

$$\frac{\partial u}{\partial t}(x, t) = V(x)u(x, t) + Qu(\cdot, t) \tag{11}$$

with initial condition

$$u(x, 0) = f(x). \tag{12}$$

If X_t is the Markov process corresponding to Q, a representation for the solution is derived in [18], [16]. Under appropriate conditions it takes the form

$$u(x, t) = E^x[M_{0t} f(X_t)], \tag{13}$$

where M_{st}, $0 \leqslant s \leqslant t < \infty$ are operator valued random variables solving

$$\frac{dM_{st}}{ds} = -V(X_s)M_{st}, \qquad 0 \leqslant s \leqslant t, \qquad M_{tt} = I. \tag{14}$$

When (14) has a unique solution the operators satisfy

$$M_{t_1 t_2} M_{t_2 t_3} = M_{t_1 t_3}, \qquad t_1 \leqslant t_2 \leqslant t_3, \tag{15}$$

and such a family of random operators is known as a *multiplicative operator valued functional*. For details about this concept see Pinsky [30]. Numerous examples and applications are given in Hersh and Papanicolaou [18]. Two of the simplest should be mentioned here.

Let X be a Markov process with homogeneous transition probabilities on a finite state space $S = \{1, 2, \ldots, n\}$, and $Q = (q_{ij})$ as at the beginning of this section. Writing $u_i(\varepsilon)$ for $u(i, t)$ and V_i for $V(i)$, (11) becomes

$$\frac{du_i}{dt} = V_i u_i + \sum_{i=1}^{n} q_{ij} u_j.$$

(By specializing the V_i one is back to the first example discussed.) If each V_i is the generator of a strongly continuous contraction semigroup $T_t^{(i)}$ on B (not necessarily coming from a Markov process), M_{st} can be given explicitly. If $\tau_1 < \tau_2 < \cdots < \tau_n$ are the successive jump times of X up to time t,

$$M_{st} = T_{\tau_1 - s}^{X_s} T_{\tau_2 - \tau_1}^{X_{\tau_1}} \cdots T_{t - \tau_N}^{X_{\tau}}.$$

Finally let $S = R^d$, $B = R^1$, Q a canonical diffusion operator, $V(x)$ a real valued continuous function bounded from above. In this case $u(x, t)$ is real valued. Then

$$M_{st} = \exp\left\{ \int_s^t V(X(\tau))\, d\tau \right\},$$

and one obtains that the equation

$$\frac{\partial u}{\partial t}(x, t) = V(x)u(x, t) + Lu(x, t), \qquad u(x, 0) = f(x) \quad (16)$$

is solved by

$$u(x, t) = E^x\left[\exp\left\{ \int_0^t V(X_s)\, ds \right\} f(X_t) \right] \quad (17)$$

which is the well-known Feynman-Kac *formula*.

(B) Non-linear Diffusion. By introducing into diffusion processes a suitable branching mechanism, one is led to nonlinear diffusion operators.

Brownian motion in R^1 can be modified to obtain the *branching Brownian motion* as follows: a particle starts at x at time 0 and performs a Brownian motion up to a random time τ, where τ is independent of X and exponentially distributed, $P[\tau > t] = e^{-t}$. At time τ, the particle splits in two, the two new particles continuing along independent Brownian paths from X_τ. Each of the new particles now continues, up to an exponentially distributed splitting time, which is independent of the motion of all particles and all other splitting times, and then the particle splits in two, etc. So at time t a random number N_t of particles $X_t^{(1)}, \ldots, X_t^{(N_t)}$ exist. As usual P^x will denote the probability measure corresponding to the process started at x.

Let ϕ be a bounded measurable function from R^1 to R^1. The function

$$u(x,t) = E^x \left[\phi\left(X_t^{(1)} \right) \cdots \phi\left(X_t^{(N_t)} \right) \right] \tag{1}$$

satisfies the nonlinear diffusion equation

$$\frac{\partial u}{\partial t} = \frac{1}{2} \frac{\partial^2 u}{\partial x^2} + u^2 - u \tag{2}$$

with initial condition

$$u(x,0) = \phi(x). \tag{3}$$

To see this let again

$$T_s g(x) = \int p(s, x, y) g(y) \, dy,$$

where $p(s, x, y)$ is the transition probability density for Brownian motion.

Let h be a small positive number ultimately tending to zero. Since the probability of more than one split by time h is $o(h)$,

$$u(x, t + h) = P^x[\tau > h] T_h u(x)$$

$$+ \int_0^h P^x[\tau \in ds] T_s [T_{h-s} u]^2 \, ds + o(h)$$

$$= (1 - h) T_h u + \int_0^h e^{-s} T_s [T_{h-s} u]^2 \, ds + o(h),$$

and so

$$\frac{1}{h} [u(x, t + h) - u(x, t)] = \frac{1}{h}(T_h - I)u - T_h u + (T_h u)^2 + o(1),$$

and letting $h \to 0$ gives (2).

It is a remarkable fact that this simplest of all branching diffusion models leads to one of the oldest and most studied nonlinear diffusion equations: (2) is the Kolmogorow-Petrowsky-Piskounov equation studied originally in [23]. (In [23] $u - u^2$ appears in place of $u^2 - u$, but this is equivalent if one changes variables, letting $v = -u$.) McKean showed in [27] that the representation (1) could be used effectively, not only to simplify the proofs of [23], but to obtain improved results. Begin with the initial condition

$$\phi(x) = \begin{cases} 1 & x > 0, \\ 0 & x \leqslant 0. \end{cases}$$

In this case

$$u(x, t) = P^x \left[\min_{i \leqslant N(t)} X_t^{(i)} > 0 \right] = P^0 \left[\max_{i \leqslant N(t)} X_t^{(i)} \leqslant x \right], \qquad (4)$$

the first identity following from (1), whilst the second follows from the translation invariance and symmetry of Brownian motion. Call a function $w_c(t)$ a *wave solution* of (2) *traveling at speed c* if it is a probability distribution function such that $w_c(x, t) \equiv w_c(x - ct)$ satisfies (2). Equivalently, $w_c(t)$ increases from $w(-\infty) = 0$ to $w(\infty) = 1$ and satisfies

$$\tfrac{1}{2} w_c'' + c w_c' + w_c^2 - w_c = 0. \qquad (5)$$

An examination of this differential equation reveals that a solution satisfying the given side conditions will exist if and only if $c \geqslant \sqrt{2}$, and for such c the solution is unique up to translation.

Let $m(t)$ be defined by $u(m(t), t) = \tfrac{1}{2}$, so that $m(t)$ is the median of the distribution function of

$$M_t = \max_{i \leqslant N_t} X_t^{(i)}$$

under P^0. In [23] it is shown that $m(t) \sim \sqrt{2}\, t$ and that

$$\lim_{t \to \infty} u(t, x + m(t)) = w_{\sqrt{2}}(x). \qquad (6)$$

So $M_t - m(t)$ has a limiting distribution, which is identified as the wave solution $w_{\sqrt{2}}$, normalized so that $w_{\sqrt{2}}(0) = \frac{1}{2}$. In fact McKean was able to obtain more precise information about the asymptotic behavior of $m(t)$. A much more delicate investigation of the branching behavior of the probabilistic model by M. Bramson [5] gives the estimate

$$m(t) = \sqrt{2}\, t - 3 \cdot 2^{-3/2} \log t + O(1). \tag{7}$$

In order to make the limit theorem fully satisfying, an estimate of $m(t)$ up to $o(1)$ terms should be obtained.[5]

The asymptotic behavior of u when other initial functions ϕ are specified in (3) has also been studied. McKean [27] (see the Correction) used probabilistic arguments to obtain a new result on this problem. The result is as follows:

> Let a and b be constants such that $a > 0$, $0 < b < \sqrt{2}$ and set $c = b^{-1} + b/2$, and suppose u solves (2), (3) and $e^{bx}(1 - \phi(x)) \to a$ as $x \to \infty$; then $u(x + ct, ct) \to w_c(x)$.

A key observation is that on setting

$$Z_t = e^{-t} \sum_{i=1}^{n_t} \exp\left\{ -bX_t^{(i)} - \frac{1}{2}b^2 t \right\}$$

and taking $\mathcal{F}_t$ to be the least σ-field with respect to which Z_t is measurable, one finds that $(Z_t, \mathcal{F}_t, 0 \le t < \infty)$ is a martingale with respect to P^0. From the relation (1) between u and ϕ and the assumptions on ϕ, one quickly obtains that $u(x + ct, t)$ is bounded above and below by terms of the form

$$E^0\left[\exp\{ -(1 - o(1))ae^{-bx}Z_t \} \right] + o(1).$$

By the martingale convergence theorem, as $t \to \infty$ Z_t converges a.s.

---

[5] This improvement, and many other results, have in the meantime been obtained by M. Bramson. The work is forthcoming as an AMS Memoir.

to a limit Z_∞ and so

$$\lim_{t \to \infty} u(x + ct, t) = E^0 \left[\exp\{- ae^{-bx} Z_\infty\} \right]. \tag{8}$$

For $c > \sqrt{2}$, examination of (5) reveals that

$$e^{bx} \left[1 - w_c(x) \right] \to a' > 0 \quad \text{as} \quad x \to \infty.$$

Now w_c is determined only up to a translation, and by choosing the appropriate translation one can arrange $a' = a$. Hence, repeating the argument with w_c in place of ϕ reveals

$$w_c(x + ct, t) \equiv w_c(x) = E^0 \left[\exp\{- ae^{-bx} Z_\infty\} \right]$$

and so the right side of (8) has been identified.

It is natural to consider in place of (2) the more general equation

$$\frac{\partial u}{\partial t} = \frac{1}{2} \frac{\partial^2 u}{\partial x^2} + f(u)$$

for suitable functions f. Such problems have been investigated intensively. The reader is referred to Aronson and Weinberger [2], [3].

REFERENCES

1. L. Arnold, *Stochastic Differential Equations: Theory and Applications*, Wiley-Interscience, New York, 1973.
2. D. G. Aronson, H. F. Weinberger, "Nonlinear diffusion in population genetics, combustion, and nerve propagation," *Partial Differential Equations and Related Topics*, ed. J. A. Goldstein, *Lecture Notes in Mathematics*, 446, Springer, New York, 1975, pp. 5–49.
3. ———, "Multidimensional nonlinear diffusion arising in population genetics," *Adv. in Math.*, **30** (1978), 37–76.
4. R. M. Blumenthal, R. K. Getoor, *Markov Processes and Potential Theory*, Academic Press, New York, 1968.
5. M. D. Bramson, "Maximal displacement of Brownian motion," *Comm. Pure Appl. Math.*, **31** (1978), 531–581.
6. K. K. Chung, P. Erdős, T. Sirao, "On the Lipschitz condition for Brownian motion," *J. Math. Soc. Japan*, **11** (1959), 263–274.

7. E. Conway, "On the total variation of solutions of parabolic equations," *Indiana Univ. Math. J.*, **21** (1971), 493–503.

8. A. Devinatz, A. Friedman, "The asymptotic behavior of the solution of a singularly perturbed Dirichlet problem," *Indiana Univ. Math. J.*, **27** (1978), 143–159.

9. J. Doob, "A probability approach to the heat equation," *Trans. Amer. Math. Soc.*, **80** (1955), 216–280.

10. A. Dvoretzky, P. Erdős, "Some problems on random walk in space," *Second Berkeley Symp. on Statistics and Probability* (1951), 353–368.

11. E. B. Dynkin, *Markov Processes*, Springer-Verlag, Berlin, 1965.

12. E. B. Dynkin, A. A. Yushkevitch, *Markov Processes, Theorems and Problems*, Plenum Press, New York, 1969.

13. A. Friedman, *Stochastic Differential Equations and Applications*, Academic Press, New York, 1975.

14. R. J. Griego, R. Hersh, "Random evolutions, Markov chains, and systems of partial differential equations," *Trans Amer. Math. Soc.*, **156** (1971), 405–418.

15. D. C. Heath, "Probabilistic analysis of hyperbolic systems of partial differential equations," thesis, Univ. of Illinois (1969).

16. R. Hersh, "Theory of random evolutions with applications to partial differential equations," *Rocky Mountain J. Math.*, **4** (1974), 443–476.

17. ———, "Stochastic solutions of hyperbolic systems," *Partial Differential Equations and Related Topics*, Lecture Notes in Mathematics, **446**, Springer, New York, 1975, pp. 283–300.

18. R. Hersh, G. Papanicolaou, "Noncommuting random evolutions, and an operator-valued Feynman-Kac formula," *Comm. Pure Appl. Math.*, **25** (1972), 337–367.

19. N. Ikeda, S. Nakao, Y. Yamato, "A class of approximations of Brownian motion," *Res. Inst. Math. Sci.*, Kyoto Univ., **13** (1977), 285–300.

20. K. Ito, "Stochastic differentials," *Lecture Notes in Mathematics*, **294**, Springer-Verlag, Berlin (1972), 1–7.

21. N. C. Jain, S. J. Taylor, "Local asymptotic laws for Brownian motion," *Ann. Probab.*, **1** (1973), 527–549.

22. N. Kôno, "The exact Hausdorff measure of irregularity points for a Brownian path," *Z. f. Wahrscheinlichkeitsrechnung*, **40** (1977), 257–289.

23. A. Kolmogorov, I. Petrowsky, N. Piscounov, "Étude de l'équation de la diffusion avec croissance de la quantité de la matière et son application à un problème biologique," *Moscow Univ. Math. Bull.*, **1**, sec. A (1937), 1–37.

24. J. Lamperti, "Criteria for the recurrence or transcience of stochastic processes," *J. Math. Anal. Appl.*, **1** (1960), 314–330.

25. E. M. Landis, "Necessary and sufficient conditions for regularity of a boundary point in the Dirichlet problem for the heat conduction equation," *Soviet Math. Dokl.*, **10** (1969), 380–384.

26. M. Loève, *Probability Theory*, 4th ed., Springer-Verlag, New York, 1977.

27. H. P. McKean, "Application of Brownian motion to the equation of Kolmogorow-Pitrowsky-Piscounov," *Comm. Pure Appl. Math.*, **28** (1975), 323–331. Correction in **29** (1976), 553–554.

28. N. Meyers, J. Serrin, "The exterior Dirichlet problem for second order elliptic partial differential equations," *J. Math. and Mech.*, **9** (1960), 513–538.
29. P. Ney, F. Spitzer, "The martin boundary for random walk," *Trans. Amer. Math. Soc.*, **121** (1966), 116–132.
30. M. Pinsky, "Multiplicative operator functionals and their asymptotic properties," *Adv. in Appl. Probab.*, **3** (1974), 1–97.
31. S. Port, C. Stone, "Classical potential theory and Brownian motion," *6th Berkeley Symposium on Math. Statistics and Probability*, vol. **3** (1970).
32. ———, "Logarithmic potentials and planar Brownian motion," *6th Berkeley Symposium on Math. Statistics and Probability*, vol. **3** (1970).
33. M. Rao, *Brownian Motion and Classical Potential Theory*, Aarhus, 1976.
34. F. Spitzer, "Some theorems concerning two-dimensional Brownian motion," *Trans. Amer. Math. Soc.*, **87** (1958), 187–197.
35. ———, *Principles of Random Walk*, 2nd ed., Springer-Verlag, New York (1976).
36. D. Stroock, S. R. S. Varadhan, "On the support of diffusion processes, with applications to the strong maximum principle," *6th Berkeley Symposium on Probability and Statistics*, **3** (1972), 333–360.
37. ———, "On degenerate elliptic-parabolic operators of second order and their associated diffusions," *Comm. Pure Appl. Math.*, **25** (1972), 651–713.
38. ———, *Multidimensional Diffusion Processes*, Springer-Verlag, Berlin, 1979.
39. A. D. Ventcel, M. J. Freidlin, "On small random perturbations of dynamical systems," *Russian Math. Surveys*, **25** (1970), 1–56.
40. S. R. S. Varadhan, "Diffusion processes in a small time interval," *Comm. Pure Appl. Math.*, **20** (1967), 659–685.

October 1979

GAUSSIAN BEAMS AND THE PROPAGATION OF SINGULARITIES

James Ralston

1. INTRODUCTION

This article deals with linear hyperbolic partial differential equations. Hyperbolic equations govern wave motion, and one recognizes waves by the intervals of stillness that precede and follow their passage. This property of waves corresponds to a distinctive feature of these equations: linear hyperbolic equations admit solutions which are concentrated near certain curves in space-time. The "gaussian beams" of the title are waves whose amplitudes at any given time are nearly gaussian distributions with small variances. These waves propagate in an especially simple fashion and it is possible to construct them rather explicitly. Moreover, one can treat them as fundamental building blocks of wave motion and use them to study general solutions of the partial differential equations.

A second distinctive feature of hyperbolic equations—in comparison with elliptic equations—is that they admit solutions which are not infinitely differentiable.[1] The problem of "propagation of singularities" is the problem of determining where the solutions can fail to be infinitely differentiable. The term "propagation" of singularities arises because one can think of singularities at one time propagating into new singularities at a later time. As one might expect, the propagation of singularities is closely tied to the existence of highly localized solutions. In fact, a singularity at x_0 at time t_0 can go to x_1 at time t_1 only if there is a gaussian beam peaked at x_1 at time t_1 with a peak at x_0 at time t_0.

The class of strictly hyperbolic operators is defined as follows. Given a linear partial differential operator P of order m, acting on functions defined for $(x, t) \in R^n \times R$, one sets $f_\alpha = \exp(i\alpha(x \cdot \xi + t\tau))$ and

$$p_m(x, t, \xi, \tau) = \lim_{\alpha \to \infty} \alpha^{-m} [\bar{f}_\alpha P(f_\alpha)](x, t).$$

The function p_m is the principal symbol of P. P is strictly hyperbolic if the polynomial $f(s) = p_m(x, t, \xi, s)$ has m distinct real roots when $\xi \neq 0$. This article deals exclusively with scalar equations

$$Pu = f,$$

where P is strictly hyperbolic. The results carry over virtually unchanged to systems of equations as long as an analogue of strict hyperbolicity holds. However, when $p_m(x, t, \xi, s)$ has multiple real roots, the propagation of singularities is much more complicated and is not yet well understood (see Uhlmann [29], Ralston [20] and the references in these papers). An immediate consequence of strict hyperbolicity is that the coefficient of $\partial^m / \partial t^m$ in P is nonzero. Normalizing, this coefficient will be assumed to be one.

The simplest (and least interesting) strictly hyperbolic operator is $Pu = \partial u / \partial t$. Since the solutions of $Pu = 0$ are the functions independent of t, $Pu = 0$ admits solutions vanishing outside small neighborhoods of the lines (x_0, t), $t \in R$, and the singularities of u propagate along these lines.

---

[1]Throughout this article coefficients are assumed to be infinitely differentiable. Without this assumption this sentence would not make sense.

The most important strictly hyperbolic operator is

$$Pu = \frac{\partial^2 u}{\partial t^2} - \sum_{i=1}^{n} \frac{\partial^2 u}{\partial x_i^2}.$$

The equation $Pu = 0$ is the classical wave equation of physics. This equation is harder to solve and we won't exhibit localized solutions until gaussian beams have been constructed (see Section 5). However, the wave equation is not qualitatively very different from the first example. In this case there are solutions which are negligibly small away from the lines in space-time traced by

$$(x, t) = (x_0 + s\xi_0, t_0 \pm s|\xi_0|), \qquad s \in R$$

for $\xi_0 \in R^n/0$, and singularities propagate along these lines.

For a general P the curves that play the role of the distinguished lines in the examples above are projections into (x, t)-space of the null bicharacteristic curves for P. Given $(x_0, t_0, \xi_0) \in R^n \times R \times R^n/0$, $(x(s), t(s), \xi(s), \tau(s))$ is a null bicharacteristic curve through (x_0, t_0, ξ_0) if it satisfies the (hamiltonian) system of ordinary differential equations

$$\dot{x} = \frac{\partial p_m}{\partial \xi} \qquad \dot{t} = \frac{\partial p_m}{\partial \tau} \qquad \dot{\xi} = -\frac{\partial p_m}{\partial x} \qquad \dot{\tau} = -\frac{\partial p_m}{\partial t}$$

with $(x(0), t(0), \xi(0)) = (x_0, t_0, \xi_0)$ and $\tau(0)$ chosen so that $p_m(x_0, t_0, \xi_0, \tau(0)) = 0$. Note that there are m choices for $\tau(0)$ and that $p_m(x(s), t(s), \xi(s), \tau(s)) = 0$ for all s. To abbreviate the terminology, the projection of a bicharacteristic curve on (x, t)-space is called a "ray path." Briefly stated, the objective of this article is to construct nontrivial asymptotic solutions to $Pu = 0$ which are concentrated near a given ray path and to show that singularities in solutions of $Pu = f$ can only propagate along ray paths.

The first use of constructions like gaussian beams in the study of the propagation of singularities appears in Hörmander [12]. In the mathematical physics literature, the use of gaussian beams as approximate solutions to wave equations began somewhat earlier (see Babich and Buldyrev [30]). A survey of this work and more recent applications can be found in Arnaud [1], [2]. In any case one sees that the techniques that will be used here are recent developments. The results on the propagation of singularities are much older. The observation that wave equations do admit solutions with singularities was made in the nineteenth century. However, the first general theorems on the propagation of singularities are contained in the work of Courant, Lax, and Ludwig in the 1950's ([5], [13], [14]). These results were obtained by the construction of "parametrices," and we will digress a bit to describe this approach and subsequent developments.

The most natural boundary value problem for a strictly hyperbolic operator P is the Cauchy problem

$$Pu = 0 \text{ in } R^n \times R, \qquad \frac{\partial^i u}{\partial t^i}(x, t_0) = g_i(x), \qquad i = 0, \ldots, m-1.$$

This is always well-posed, and one has the linear operator

$$R(t, t_0): \left(u(x, t_0), \ldots, \frac{\partial^{m-1} u}{\partial t^{m-1}}(x, t_0) \right) \to u(x, t).$$

In [13] Lax constructed $R(t_1, t_0)$ modulo an integral operator whose kernel could be made r times continuously differentiable for any r. This made it possible to read off the propagation of singularities. Unfortunately Lax's construction was possible only for $|t_1 - t_0|$ small. In [14] Ludwig pointed out that this could be overcome by writing $R(t_1, t_0)$ as a product of $R(t_i, t_{i-1})$ with $|t_i - t_{i-1}|$ small. However, in [6] Hörmander and Duistermaat gave a general procedure for constructing fundamental solutions like $R(t_1, t_0)$ globally in space-time modulo integral operators with infinitely differentiable kernels. In addition in [11], Hörmander introduced a refinement of the notion of singularity, the "wave front set." This sharpened all previous results considerably: if u has a singularity at (x_0, t_0), this singularity will in general not propagate along all ray paths through (x_0, t_0). If you know the wave front set of the restriction of u to $t = t_0$, you can say a great deal about which paths the singularities will take. Since Hörmander's work it has become standard, whenever possible, to state results on the propagation of singularities in terms of the propagation of wave front sets. We discuss wave front sets in Section 3, and the result we give in Section 4 is stated in terms of wave front sets.

Once the Cauchy problem was understood, it was natural to consider other boundary value problems. This was done by Povzner and Suharevskii [19], Lax and Nirenberg [18], Majda and Osher [15], and M. Taylor [27]. In these problems, when a ray path strikes a boundary nontangentially, it produces a number of reflected rays. To give the flavor of these results we have carried out both the construction of gaussian beams in Section 2 and the proof of the result on propagation of singularities in Section 4 for mixed initial-boundary value problems.

The final level of complication in the propagation of singularities comes when a ray strikes a boundary tangentially. This is the problem of "grazing" and "gliding" rays, and it has not yet been completely resolved even for the classical wave equation. However, for boundaries that are either strictly convex or strictly concave with respect to ray paths, complete results have been obtained by Melrose [17], Taylor [28], Anderssen and Melrose [3, 4], and Eskin [7].

All of the results cited above were obtained either by constructing parametrices, i.e., constructing fundamental solutions like $R(t_1, t_0)$ modulo operators with smooth kernels or by ingenious nonconstructive arguments based on building operators with prescribed commutators with P that were introduced by Hörmander in [10] and [12]. The latter argument has no direct relation to gaussian beam constructions, but the approach via parametrices does. Assume that $R(t_1, t_0)$ has been constructed modulo an integral operator with a smooth kernel, i.e., assume that a parametrix for the Cauchy problem is known. Choosing $\phi \in C_0^\infty(R^n)$ with $\phi(x) = 1$ for $|x| < R$, let $u(x, t, k)$ be the solution of the Cauchy problem

$$u(x, t_0, k) = \phi(x) \exp\left(ikx \cdot \xi - \frac{k}{2}|x - x_0|^2 \right)$$

$$\frac{\partial^i u}{\partial t^i}(x, t_0, k) = 0, \quad i = 1, \ldots, m-1, \quad \xi \in R^n/0, \quad k \in R_+.$$

Since an operator with an infinitely differentiable kernel applied to these highly oscillatory initial data produces a function which tends to zero faster than any inverse power of k, it will be possible, in principle, to determine $u(x, t_1, k)$ up to an error which decreases rapidly as $k \to +\infty$. From the form of the parametrix it will also be possible to see that as $k \to +\infty$ $u(x, t, k)$ becomes concentrated near ray paths corresponding to null bicharacteristics with initial data (x_0, t_0, ξ).

The gaussian beam construction is simply a more direct method for determining $u(x, t_1, k)$ up to an error that decreases rapidly as $k \to \infty$. One postulates that $u(x, t, k)$ will have a certain dependence on k and then constructs $v(x, t, k; x_0, \xi)$ of this form so that for $i = 0, \dots, m - 1$,

$$\left\| \left\| \frac{\partial^i u}{\partial t^i}(x, t_0, k) - \frac{\partial^i v}{\partial t^i}(x, t_0, k) \right\| \right\|$$

$$= O(k^{-N}) \quad \text{and} \quad \|Pv\| = O(k^{-N}).$$

Here $\| \ \|$ and $\|\| \ \|\|$ denote suitable Sobolev norms. Since the Cauchy problem is well-posed in these norms, one has

$$\|\|u(x, t_1, k) - v(x, t_1, k)\|\| = O(k^{-N}). \tag{1.1}$$

This procedure succeeds because one is able to guess the dependence of $u(x, t, k)$ on k. Whenever this is possible, the method of gaussian beams or, more generally, geometric optics provides a direct construction of solutions concentrated near ray paths. This, in turn, as will be shown in Section 4, leads to results on the propagation of singularities. However, the method has the limitation that it introduces extraneous hypotheses to avoid cases where the dependence of u on k becomes complicated.

To recover $R(t_1, t_0)$ from the gaussian beams one can let $\hat{f}(\xi)$ denote the Fourier transform of f and form the integral operator

$$\tilde{R}(t_1, t_0)f = \int_{R^n} \left(\frac{|\xi|}{2\pi} \right)^{n/2} \int_{R^n} v(x, t_1, |\xi|; x_0, \xi/|\xi|) \hat{f}(\xi) \, dx_0 \, d\xi.$$

If the N in (1.1) is sufficiently large and f is supported in $|x| < R$, one can show $\tilde{R}(t_1, t_0)$ differs from $R(t_1, t_0)$, restricted to data $(f, 0, \ldots, 0)$, by an operator whose kernel has any given degree of smoothness. To get an operator which differs from $R(t_1, t_0)$ by an operator with an infinitely differentiable kernel, one must use the asymptotic summation method from the theory of pseudodifferential operators. As it stands the operator $\tilde{R}(t_1, t_0)$ is rather complicated. Much of the usefulness of the calculus of Fourier integral operators developed in [11] and [6] arises because it leads to a simpler expression for $R(t_1, t_0)$ modulo operators with smooth kernels, namely a sum of operators

$$R_j(t_1, t_0)f = \int_{R^n} d\theta \int_{R^n} e^{i\psi_j(x, y, \theta; t_1, t_0)} a_j(x, y, \theta; t_1, t_0) f(y) \, dy,$$

where ψ_j and a_j are well-understood. To reduce $\tilde{R}(t_1, t_0)$ to this form it would be necessary to use the calculus developed by Melin and Sjöstrand in [16].

There was not space in this article to discuss the applications of results on propagation of singularities. Even in a much longer article this would be difficult, because almost all recent results on linear hyperbolic equations are to varying degrees related to the propagation of singularities. However, here is a short list of diverse applications with recent papers:

(i) Sharp growth estimates on spectral functions of elliptic operators, Hörmander [9].

(ii) Uniform local decay estimates for solutions of hyperbolic equations, Rauch [22].

(iii) Poisson relations for general manifolds, Andersson and Melrose [4].

In Section 5 we do include an application to boundary control for the wave equation that has not appeared elsewhere.

In conclusion I should say that throughout this article I have opted for ease of exposition over generality. In Section 4 in particular the propagation of singularities theorem is strictly contained in the results of the references given earlier. Also for ease of exposi-

tion pseudodifferential operators have been entirely suppressed—which makes this a misleading introduction to current
research.

This article was written in the summer and fall of 1977. Since
then there has been considerable progress in understanding the
propagation of singularities. In particular, the articles [32], [33] of
Melrose and Sjöstrand give an essentially complete description of
the propagation of singularities for equations like the classical wave
equation, and a systematic treatment of much of the theory of
grazing rays is now available in the book of Taylor [34]. For an
introduction to recent work in the field, one should see the conference proceedings [31].

2.1. THE CONSTRUCTION OF GAUSSIAN BEAMS

In this section we will suppress the time variable t. Thus we use
$x = (x_0, x_1, \ldots, x_n)$ where $x_0 = t$ in our earlier notation. We will
consider functions u of the form

$$u(x, k) = e^{ik\psi(x)} \left(a_0(x) + \frac{a_1(x)}{k} + \cdots + \frac{a_N(x)}{k^N} \right). \quad (2.1)$$

This is the geometric optics ansatz for the dependence of u on the
positive real parameter k. Writing the strictly hyperbolic equation
we wish to solve as $Pv = 0$, our objective here is, given M, to choose
$\psi(x), a_0(x), \ldots, a_N(x)$ so that $\|Pu\| \leqslant C_M k^{-M}$, where $\| \ \|$ is a
Sobolev norm that we'll specify later.

The gaussian beam ansatz is a special case of (2.1) in which the
function u becomes concentrated on the projection of a single null
bicharacteristic as $k \to \infty$. Letting $p_m(x, \xi)$ be the principal symbol
of P, we consider a null bicharacteristic curve

$$\dot{x} = \frac{\partial p_m}{\partial \xi}(x, \xi), \quad \dot{\xi} = -\frac{\partial p_m}{\partial x}(x, \xi), \quad x(0) = y, \quad \xi(0) = \eta \neq 0,$$

where $p_m(y, \eta) = 0$ and hence $p_m(x(s), \xi(s)) = 0$ for all s. Note that, since P is strictly hyperbolic, $\dfrac{\partial p_m}{\partial \xi_0}(x(s), \xi(s)) \neq 0$ and hence $\dot{x}(s) \neq 0$. Thus $x(s)$ traces a smooth curve Γ in R^{n+1}. We will choose $\psi(x)$ so that $\psi(x(s))$ is real-valued, but Im $\dfrac{\partial^2 \psi}{\partial x_i \partial x_j}(x(s))$ is positive definite on vectors orthogonal to $\dot{x}(s)$. Thus $|\exp ik\psi(x)|$ will look like a gaussian distribution, with variance proportional to k^{-1}, on planes perpendicular to Γ. This very rapid decrease off Γ will greatly simplify the equations we need to solve in order to get $\|Pu\| \leqslant C_M k^{-M}$. In fact this estimate will follow from the vanishing of Pu on Γ to a sufficiently high order. For the construction we will only need to make the proper choices of the Taylor series of ψ and $a_1, a_2, \ldots, a_N$ along Γ.

Since $Pu = k^m p_m\left(x, \dfrac{\partial \psi}{\partial x}\right) a_0 e^{ik\psi} + O(k^{m-1})$, the first, and by far the most important, step in the construction is the choice of ψ so that $p_m(x, \partial \psi / \partial x)$ vanishes to high order on Γ. For this we prescribe $\dfrac{\partial \psi}{\partial x}(x(s)) = \xi(s)$, and compute the Taylor series of $f(x)$ $= p_m\left(x, \dfrac{\partial \psi}{\partial x}(x)\right)$ about $x(s)$: using the summation convention, we require

$$0 = \frac{\partial f}{\partial x_j} = \frac{\partial p_m}{\partial x_j} + \frac{\partial p_m}{\partial \xi_k} \frac{\partial^2 \psi}{\partial x_j \partial x_k}, \tag{2.2}$$

$$0 = \frac{\partial^2 f}{\partial x_i \partial x_j} = \frac{\partial^2 p_m}{\partial x_i \partial x_j} + \frac{\partial^2 p_m}{\partial x_j \partial \xi_k} \frac{\partial^2 \psi}{\partial x_i \partial x_k} + \frac{\partial^2 p_m}{\partial x_i \partial \xi_k} \frac{\partial^2 \psi}{\partial x_j \partial x_k}$$

$$+ \frac{\partial^2 \psi}{\partial x_i \partial x_l} \frac{\partial^2 p_m}{\partial \xi_l \partial \xi_k} \frac{\partial^2 \psi}{\partial x_j \partial x_k} + \frac{\partial p_m}{\partial \xi_k} \frac{\partial^3 \psi}{\partial x_i \partial x_k \partial x_j}. \tag{2.3}$$

Since $\dfrac{\partial p_m}{\partial x}(x(s), \xi(s)) = -\dot{\xi}(s)$ and $\dfrac{\partial p_m}{\partial \xi}(x(s), \xi(s)) = \dot{x}(s)$, (2.2) is just the compatibility condition

$$\sum_{j=0}^{n} \frac{\partial^2 \psi}{\partial x_i \partial x_j}(x(s)) \dot{x}_j(s) = \frac{d}{ds}\left(\frac{\partial \psi}{\partial x_i}(x(s))\right).$$

Note also that, if $\dot{x}(s) \neq \frac{\partial p_m}{\partial \xi}(x(s), \xi(s))$, (2.2) implies that the null space of $\left(\text{Im} \frac{\partial^2 \psi}{\partial x_i \partial x_j} \right)$ must be at least two-dimensional, and, given

$$\dot{x}(s) = \alpha \frac{\partial p_m}{\partial \xi}(x(s), \xi(s)),$$

(2.2) forces

$$\dot{\xi}(s) = -\alpha \frac{\partial p_m}{\partial x}(x(s), \xi(s)).$$

Hence, the construction we are attempting is only *possible* if $(x(s), \xi(s))$ is a (null) bicharacteristic curve.

To study (2.3) on Γ we introduce the matrices

$$(M(s))_{ij} = \frac{\partial^2 \psi}{\partial x_i \partial x_j}(x(s)),$$

$$(A(s))_{ij} = \frac{\partial^2 p_m}{\partial x_i \partial x_j}(x(s), \xi(s)),$$

$$(B(s))_{ij} = \frac{\partial^2 p_m}{\partial \xi_i \partial x_j}(x(s), \xi(s))$$

and

$$(C(s))_{ij} = \frac{\partial^2 p_m}{\partial \xi_i \partial \xi_j}(x(s), \xi(s)).$$

Then on Γ (2.3) is the matrix equation

$$0 = A + MB + B^t M + MCM + \frac{dM}{ds}. \tag{2.4}$$

The difficulty with (2.4) is that, since it is nonlinear, one has no assurance that solutions will exist for all s. However, noting that (2.4) is a Ricatti equation for the symmetric matrix M, we proceed as follows. After choosing the $n \times n$ matrix $\left(\dfrac{\partial^2 \psi}{\partial x_i \partial x_j}(x(0)) \right)$, $1 \leqslant i$, $j \leqslant n$, so that it is symmetric with a positive definite imaginary part, we choose the rest of $M(0)$ so that $M(0)\dot{x}(0) = \xi(0)$ and $M(0) = M(0)^t$. Next, extending $y^0 = \dot{x}(0)$ to a basis $y^0, \ldots, y^n$ for R^{n+1}, we define $\eta^i = M(0)y^i$. Finally we define $M(s)$ by the equations

$$M(s)y^i(s) = \eta^i(s), \qquad i = 0, \ldots, n, \tag{2.5}$$

where $(y^i(s), \eta^i(s))$ is the solution with initial data (y^i, η^i) to the linear system

$$\dot{y} = By + C\eta,$$

$$\dot{\eta} = -Ay - B^t\eta. \tag{2.6}$$

Since all the coefficients in (2.6) are real and A and C are symmetric, the flow defined by (2.6) preserves both the symplectic form

$$\sigma(\{y, \eta\}, \{\tilde{y}, \tilde{\eta}\}) = y \cdot \tilde{\eta} - \eta \cdot \tilde{y}$$

and the complexified form $\sigma_{\mathrm{C}}(v, w) = \sigma(v, \bar{w})$. Thus, if $\sum_{i=0}^n a_i y^i(s_0) = 0$ and we set $v = \sum_{i=0}^n a_i \{y^i(s_0), \eta^i(s_0)\}$,

$$0 = \sigma_{\mathrm{C}}(v, v) = -2iy \cdot \mathrm{Im}\, M(0)\bar{y}$$

where $y = \sum_{i=0}^n a_i y^i(0)$. Thus $a_i = 0$, $i = 0, \ldots, n$, and $M(s)$ is well-defined by (2.5). Since σ is preserved by (2.6), $M(s) = M^t(s)$ for all s. One checks that $(y^0(s), n^0(s)) = (\dot{x}(s), \xi(s))$, and hence the preservation of σ_{C} implies $\mathrm{Im}\, M(s)$ is positive definite on the orthogonal complement of $\dot{x}(s)$. Finally one may check directly that $M(s)$ satisfies (2.4). Thus we have found a choice for the second order derivatives of ψ along Γ which meets all our conditions. The key observation is that, since (2.6) preserves σ and σ_{C}, solutions to (2.4) with positive imaginary part on the orthogonal complement of $\dot{x}(0)$ cannot blow up.

Writing down the analogue of (2.2) and (2.3) for the third order terms in the Taylor series $f(x) = p_m\left(x, \dfrac{\partial \psi}{\partial x}(x)\right)$ one sees that the analogue of (2.4) is an inhomogeneous *linear* ordinary differential equation along Γ. Thus, no longer worried about solutions blowing up, we only need to solve the equations corresponding to

$$\frac{\partial^3 f}{\partial x_i \partial x_j \partial x_k}(x(s)) = 0, \ 1 \leqslant i \leqslant j \leqslant k$$

for

$$\frac{\partial^3 \psi}{\partial x_i \partial x_j \partial x_k}(x(s)), \ 1 \leqslant i \leqslant j \leqslant k.$$

Where the other third order partial derivatives of ψ appear in these equations, we substitute for them from the compatibility equations

$$\sum_{k=0}^{n} \frac{\partial^3 \psi}{\partial x_i \partial x_j \partial x_k}(x(s))\dot{x}_k(s) = \frac{d}{ds}\left(\frac{\partial^2 \psi}{\partial x_i \partial x_j}(x(s))\right).$$

Since the differential equations are linear, prescribing $\dfrac{\partial^3 \psi}{\partial x_i \partial x_j \partial x_k}(x(0))$, $1 \leqslant i \leqslant j \leqslant k \leqslant n$ arbitrarily, we get the third order partials of ψ on all of Γ. We employ the same procedure for the higher order partials of ψ along Γ.

To summarize we can solve $p_m(x, \partial \psi / \partial x) = 0$ to arbitrary finite order on Γ provided we choose $\dfrac{\partial \psi}{\partial x}(x(s)) = \xi(s)$ and

$$\mathrm{Im}\left(\frac{\partial^2 \psi}{\partial x_i \partial x_j}(x(0))\right)_{1 \leqslant i, j \leqslant n}$$

positive definite. Aside from these requirements, the Taylor series at $x(0)$ of the restriction of ψ to $x_0 = x_0(0)$ can be prescribed arbitrarily. Actually by a slight extension of Borel's theorem on the existence of C^∞ functions with prescribed Taylor series, we could make $p_m(x, \partial \psi / \partial x)$ vanish to infinite order on Γ. However, the fact we will need later is that we can choose ψ depending smoothly on $y = x(0)$, $\eta = \xi(0)$, and its Taylor series at y up to order R so that the Taylor series of $p_m(x, \partial \psi / \partial x)$ vanishes to order R on $\Gamma \cap \{|x_0| < T\}$. This is clear from the preceding discussion.

To determine the Taylor series of $a_0, \ldots, a_N$ along Γ, one proceeds as follows. Define the coefficients $c_s(x)$ by

$$Pu = \left(\sum_{s=-m}^{N} c_s(x) k^{-s} \right) e^{ik\psi}.$$

Then, using the summation convention,

$$c_{-m}(x) = p_m\left(x, \frac{\partial \psi}{\partial x}\right) a_0,$$

$$c_{-m+1}(x) = \frac{1}{i}\left(\frac{\partial p_m}{\partial \xi_j}\left(x, \frac{\partial \psi}{\partial x}\right) \frac{\partial a_0}{\partial x_j} \right)$$

$$+ \left(\frac{1}{2i} \frac{\partial^2 p_m}{\partial \xi_j \partial \xi_k}\left(x, \frac{\partial \psi}{\partial x}\right) \frac{\partial^2 \psi}{\partial x_j \partial x_k} + p_{m-1}\left(x, \frac{\partial \psi}{\partial x}\right) \right) a_0$$

$$+ p_m\left(x, \frac{\partial \psi}{\partial x}\right) a_1$$

$$\triangleq La_0 + p_m\left(x, \frac{\partial \psi}{\partial x}\right) a_1, \quad \text{and}$$

$$c_{-m+r+1}(x) = La_r + p_m\left(x, \frac{\partial \psi}{\partial x}\right) a_{r+1} + g_r, \qquad r = 1, \ldots, N+m,$$

$$(2.7)$$

where g_r is a complicated function of ψ, $a_0, \ldots, a_{r-1}$ and their derivatives, and we have defined $a_r \equiv 0$ for $r > N$. One sees immediately that, given $p_m(x, \partial \psi / \partial x)$ vanishes to order R on Γ, we can choose the Taylor series of a_0 on Γ up to order $R - 2$ on Γ so that C_{-m+1} vanishes to order $R - 2$ on Γ. Moreover, a_0 can be chosen as a smooth function of its Taylor series up to order $R - 2$ at $x(0)$ and the Taylor series of ψ at $x(0)$ up to order R. Passing to the higher order equations $c_{-m+s}(x) = 0$, we see that we can choose a_{s-1} so that $c_{-m+s}(x)$ vanishes on Γ to order $R - 2s$ and a_{s-1} is a smooth function of its Taylor series up to order $R - 2s$ at $x(0)$, the Taylor series of a_r, $r = 0, \ldots, s - 2$, at $x(0)$ up to order $R - 2r$ and the Taylor series of ψ up to order R. The number of terms, N, in the sum (2.1) is determined by the condition $2 > R - 2N \geqslant 0$.

The final step in the construction of u is to multiply the function given by (2.1) by a k-independent smooth function which is identically one on a neighborhood of Γ, but which vanishes outside a small neighborhood $\mathcal{O}$ of Γ. Given T we always choose $\mathcal{O}$ so that $\operatorname{Im} \psi(x) \geqslant cd^2(x)$ for $x \in \mathcal{O} \cap \{|x_0| \leqslant T\}$ where $d(x)$ is the distance from x to Γ and $c > 0$.

Now we are ready to estimate $\|Pu\|$. Our estimates are immediate consequences of the following elementary lemma.

LEMMA (2.8). *Assume $c(x)$ vanishes to order $S - 1$ on Γ and the intersection of support c with $|x_0| \leqslant T$ is compact. Assume further that $\operatorname{Im} \psi(x) \geqslant ad^2(x)$ on the support of c for $|x_0| \leqslant T$. Then*

$$\int_{|x_0| \leqslant T} |ce^{ik\psi}|^2 \, dx \leqslant Ck^{-S-n/2}.$$

Proof. In a neighborhood of $\Gamma \cap |x_0| \leqslant T$ we may introduce (k-independent) coordinates z so that Γ is traced by $z_0 = s$, $z_i = 0$ $i = 1$, n and $d^2(x(z)) \geqslant z_1^2 + \cdots + z_n^2$. Then, introducing the stretched coordinates y, $y_0 = z_0$, $y_i = k^{1/2} z_i$, $i = 1, \ldots, n$, and changing from x to y variables in the integral, we see that the new integrand is bounded by

$$Ck^{-S-n/2} \big| k^{S/2} c\big(x\big(y_0, k^{-1/2} y' \big) \big) \big|^2 \exp\{ -2a|y'|^2 \},$$

where $y = (y_0, y')$. Since c vanishes to order $S - 1$ on Γ,

$$\big| k^{S/2} c\big(x\big(y_0, k^{-1/2} y' \big) \big) \big|$$

is bounded as $k \to \infty$. The estimate of the lemma follows.

To estimate Sobolev norms of Pu we note that differentiation of Pu will either multiply the functions $c_i(x)$, $i = -m,\ldots,N$, by k or decrease the order that they vanish on Γ by at most one. Thus, letting $\| \ \|_s$ denote the Sobolev s-norm on $|x_0| \leqslant T$, we see by repeated application of the lemma

$$\|Pu\|_s \leqslant Ck^{m+s-(R+1)/2-n/4}.$$

Moreover, the constant is uniform in $y = x(0)$, $\eta = \xi(0)$, the Taylor series of ψ at $x(0)$ up to order R, and the Taylor series of a_s up to order $R - 2s$, provided all these data lie in a bounded set and one has a uniform bound on the positivity of

$$\left(\text{Im} \frac{\partial^2 \psi}{\partial x_i \partial x_j}(x(0)) \right)_{1 \leqslant i, j \leqslant n}$$

2.2. REFLECTIONS

In this subsection we will consider boundary value problems. To do this we replace the domain R^{n+1} by $D = R \times \Omega$, where Ω is a domain in R^n with smooth boundary. On $\partial D = R \times \partial \Omega$ we impose the boundary conditions $B_i u = 0$, $i = 1,\ldots,l$, where B_i is a linear differential operator. If the ray path $x(s)$ strikes ∂D at $x(s_0)$, then the preceding constructions must be modified so that the gaussian beams will satisfy the boundary conditions. To carry out this modification, we must make two assumptions. Letting ν denote the inner unit normal to ∂D at $x(s_0)$, we assume that $p_m(x(s_0), \xi(s_0) + t\nu) = 0$ has m distinct roots t in the complex numbers. Since this condition reduces to $\nu \cdot \dot{x}(s_0) \neq 0$ when one considers the classical wave equation, this assumption is known as the "nongrazing hypothesis." Secondly, given that the roots are distinct, we let k be the number of real ones. The real roots τ for which

$$\left[\left(\nu \cdot \frac{\partial p_m}{\partial \xi} \right) \left(\frac{\partial p_m}{\partial \xi_0} \right) \right](x(s_0), \xi(s_0) + \tau\nu) > 0$$

will be labelled τ_i, $i = 1, \ldots, k_0$. Since the complex roots occur in conjugate pairs, we can label them so that $\operatorname{Im}\tau_i > 0$, $i = k_0 + 1, \ldots,$ $k_0 + (m - k)/2$. Then, letting $b_i(x, \xi)$ be the principal symbol of B_i, we assume that the matrix b defined by

$$(b)_{ij} = b_i\big(x(s_0), \xi(s_0) + \tau_j \nu\big)$$

$i = 1, \ldots, l; j = 1, \ldots, k_0 + (m - k)/2$, has rank l. We will see that if b has a nontrivial null space and rank l, the mixed boundary-initial value problem

$$Pu = 0 \text{ on } R_+ \times \Omega,$$

$$\frac{\partial^s u}{\partial t^s}(0, x') = g_s(x') \quad \text{on } \Omega, \quad s = 0, \ldots, m-1, \qquad (2.9)$$

$$B_i u = 0 \text{ on } R_+ \times \partial\Omega, \qquad i = 1, \ldots, l,$$

is not well-posed. Thus we may as well assume $l = k_0 + (m - k)/2$. With this addition, given the "nongrazing" hypothesis, the second hypothesis is implied by Sakamoto's criterion for (2.9) to be well-posed [25]. However, it is worth pointing out that whether or not B has rank l can depend on how one writes the boundary conditions.

Given these hypotheses, the construction of a "reflected" gaussian beam satisfying the boundary conditions to high order in k^{-1} is not difficult. We replace the ansatz (2.1) by

$$u = e^{ik\psi}\left(a_0 + \cdots + \frac{a_N}{k^N}\right) + \sum_{j=1}^{l} e^{ik\psi_j}\left(a_1^j + \cdots + \frac{a_N^j}{k^N}\right). \quad (2.10)$$

For $j = 1, \ldots, k_0$, ψ_j will be built by the method described in Section 1 with $x(s_0)$ playing the role of $x(0)$ and $\xi(s_0) + \tau_j \nu$ in place of $\xi(0)$. Note that $\xi(s_0) + \tau_j \nu \neq 0$ by the nongrazing hypothesis. To determine initial data for ψ_j we require $\psi_j = \psi$ on ∂D. Since the nongrazing hypothesis implies

$$\nu \cdot \frac{\partial p_m}{\partial \xi}\big(x(s_0), \xi(s_0) + \tau_j \nu\big) \neq 0,$$

the Taylor series of ψ_j at $x(s_0)$ is determined by this requirement and the compatibility conditions

$$\sum_{i=0}^{n} \frac{\partial^{|\alpha|+1}\psi_j}{\partial x^\alpha \partial x_i}\big(x(s_0)\big)\frac{\partial p_m}{\partial \xi_i}\big(x(s_0), \xi(s_0)+\tau_j\nu\big)$$

$$= \frac{d}{ds}\frac{\partial^{|\alpha|}\psi_j}{\partial x^\alpha}\big(\underline{x}(s)\big)\bigg|_{s=0},$$

$$(2.11)$$

where $(\underline{x}(s), \underline{\xi}(s))$ is the bicharacteristic with initial data $(x(s_0), \xi(s_0)+\underline{\tau}_j\nu)$. Moreover, the nongrazing hypothesis also implies $\nu \cdot \dot{x}(s_0) \neq 0$ so that $\mathrm{Im}\dfrac{\partial^2 \psi}{\partial x_i \partial x_k}(x(s_0))$ is positive definite on the tangent plane to $R \times \partial D$ at $x(s_0)$. Thus $\mathrm{Im}\dfrac{\partial^2 \psi_j}{\partial x_i \partial x_k}(x(s_0))$ is also positive definite on this plane and, hence, on all planes transverse to $\underline{\dot{x}}(s_0)$. This shows that we can employ the method of Section 2.1 to choose ψ_j so that $p_m(x, \partial \psi_j/\partial x)$ vanishes to order R on the curve Γ_j traced by $\underline{x}(s)$ with $\mathrm{Im}\dfrac{\partial \psi_j}{\partial x_i \partial x_j}(\underline{x}(s))$ positive definite on the orthogonal complement of $\underline{\dot{x}}(s)$.

The construction of ψ_j, $j = k_0+1,\ldots,k_0+(m-k)/2$, is somewhat different. Here it will suffice to determine the Taylor series of ψ_j at $x(s_0)$. As before we require $\psi = \psi_j$ on $R \times \partial D$, and, again the nongrazing hypothesis implies

$$\nu \cdot \frac{\partial p_m}{\partial \xi}\big(x(s_0), \xi(s_0)+\tau_j\nu\big) \neq 0,$$

and we specify $\dfrac{\partial \psi_j}{\partial x}(x(s_0))$ by

$$\frac{\partial \psi_j}{\partial x}\big(x(s_0)\big) = \xi(s_0)+\tau_j\nu.$$

Since $\dfrac{\partial p}{\partial \xi}(x(s_0), \xi(s_0) + t_j \nu)$ is complex, it does not make sense to speak of $(\underline{x}(s), \underline{\xi}(s))$, but we can replace (2.11) by

$$0 = \sum_{i=0}^{n} \frac{\partial^{|\alpha|+1} \psi_j}{\partial x^\alpha \partial x_i}(x(s_0)) \frac{\partial p_m}{\partial \xi_i}(x(s_0)), \frac{\partial \psi_j}{\partial x}(x(s_0)) + q_\alpha(x(s_0)),$$

$$(2.12)$$

where q_α is the term, depending only on $\dfrac{\partial^{|\beta|} \psi_j}{\partial x^\beta}(x(s_0))$, $|\beta| \leqslant |\alpha|$, determined by requiring that $p_m\left(x, \dfrac{\partial \psi_j}{\partial x}(x) \right)$ vanish to order $|\alpha|$ at $x(s_0)$ as in (2.2) and (2.3). Then (2.12) and the requirement $\psi = \psi_j$ determine the Taylor series of ψ_j at $x(s_0)$ uniquely.

In our construction it will be essential that for $j > k_0$,

$$\operatorname{Im} \psi_j(x) \geqslant c|x - x(s_0)|^2, \ c > 0,$$

for x in a neighborhood of $x(s_0)$ in D. To see how this can be arranged, it is convenient to assume that the coordinates near $x(s_0)$ are such that D is defined by $x_n > 0$, and $x(s_0) = 0$. Then by our preceding conditions

$$\operatorname{Im} \frac{\partial \psi_j}{\partial x}(x(s_0)) = (\operatorname{Im} \tau_j)\nu = (0, \dots, 0, \operatorname{Im} \tau_j)$$

and

$$\operatorname{Im} \frac{\partial^2 \psi_j}{\partial x_i \partial x_k}(x(s_0)) = \operatorname{Im} \frac{\partial^2 \psi}{\partial x_i \partial x_k}(x(s_0)), \quad 0 \leqslant i, k \leqslant n-1.$$

Letting $x = (x', x_n)$ and $(A)_{ij} = \operatorname{Im} \dfrac{\partial^2 \psi}{\partial x_i \partial x_j}$, $0 \leqslant i, j \leqslant n-1$, we have

$$\operatorname{Im} \psi_j = (\operatorname{Im} \tau_j)x_n + \frac{1}{2}x' \cdot Ax' + x_n \cdot bx' + O(x_n^2)$$

$$+ O(|x - x(s_0)|^3).$$

Since A is positive definite and $\text{Im}\,\tau_j > 0$, one sees immediately that for $|x - x(s_0)|$ sufficiently small, we have

$$\text{Im}\,\psi_j(x) \geqslant c|x - x(s_0)|^2,$$

when $x_n \geqslant 0$.

The final step in this construction is the determination of the coefficients $a_0^j, \ldots, a_N^j$. Once the Taylor series at $x = x(s_0)$ of the restrictions of these coefficients to ∂D have been determined one has only to use the methods of Section 2.1 for $j = 1, \ldots, k_0$ and the analogue of (2.11) for $j = k_0 + 1, \ldots, k_0 + (m-k)/2$, to choose $a_0^j, \ldots, a_N^j$ so that $c_{-m+s}^j(x)$ vanishes on Γ_j to order $R - 2s$ for $j = 1, \ldots, k_0$ and $c_{-m+s}^j(x)$ vanishes to order $R - 2s$ at $x(s_0)$ for $j = k_0 + 1, \ldots, k_0 + (m-k)/2$. To determine the needed Taylor series we consider the boundary conditions. Since $\psi = \psi_j$ on ∂D, substituting from (2.10) and assuming $B_j(x, D)$ is order m_j,

$$B_j u = \sum_{r=-m_j}^{N} d_r^j(x) k^{-r} e^{ik\psi}, \qquad j = 1, \ldots, l,$$

where

$$d_{m_j}^j = b_j\left(x, \frac{\partial \psi}{\partial x}\right) a_0 + b_j\left(x, \frac{\partial \psi_1}{\partial x}\right) a_0^1 + \cdots + b_j\left(x, \frac{\partial \psi_l}{\partial x}\right) a_0^l,$$

and

$$d_{m_j+s}^j = b_j\left(x, \frac{\partial \psi}{\partial x}\right) a_s + b_j\left(x, \frac{\partial \psi_1}{\partial x}\right) a_s^1 + \cdots + b_j\left(x, \frac{\partial \psi_l}{\partial x}\right) a_s^l + g_{js}$$

where g_{js} is a complicated function of $a_r, a_r^1, \ldots, a_r^l$, $r = 0, \ldots, s-1$, and their derivatives. Given our hypothesis that the $l \times l$ matrix $(b_j(x, (\partial \psi_k/\partial x))$ is invertible, one sees that the condition $d_{m_j+s}^j$ vanishes at $x(s_0)$ to order $R - 2s$, $j = 1, \ldots, l$, determines the Taylor series of a_s^j up to order $R - 2s$, $j = 1, \ldots, l$. This completes the construction of u. We still have

$$\|Pu\|_s \leqslant Ck^{m+s-(R+1)/2-n/4};$$

in fact a trivial modification of the lemma in Section 2.1 shows that the contributions from

$$e^{i\psi_j}\left(a_1^j + \cdots + \frac{a_N^j}{k^N}\right), j = k_0 + 1, \ldots, k_0 + (m - k)/2$$

are actually of lower order in k. Moreover, if we let $||| \ |||_{s'}$ denote the Sobolev norm on $\partial D \times R \cap |x_0| \leqslant T$ of order s',

$$||| B_i u |||_{s'} \leqslant Ck^{m_i + s' - (R+1)/2 - n/4}.$$

It is also true that all constants are uniform for $y = x(0)$ and $\eta = \xi(0)$ in a neighborhood of an admissible value, though now this neighborhood may have to be small to avoid grazing. The constants are uniform in the Taylor series of a and ψ at $x(0)$ exactly as in Section 1.1. It is also clear that the construction is well behaved for a finite number of nongrazing reflections.

At this point the astute reader will note that we have not used the hypothesis

$$\left(\nu \cdot \frac{\partial p_m}{\partial \xi}\right)\left(\frac{\partial p_m}{\partial \xi_0}\right)\Bigg|_{(x(s_0), \xi(s_0) + \tau_j \nu)} > 0, \qquad j = 1, \ldots, k_0.$$

Indeed, the preceding construction can be done without this hypothesis (and the propagation of singularities results in [15] and [18] are given without it). However, we want the terms added in (2.10) to the u of (2.1) as corrections for reflection to be supported in $x_0 > x_0(s_0) - \varepsilon$ so that the corrections made when $x_0 > 0$ will not alter the Cauchy data of u at $x_0 = 0$. The condition

$$\left(\nu \cdot \frac{\partial p_m}{\partial \xi}\right)\left(\frac{\partial p_m}{\partial \xi_0}\right)\Bigg|_{(x(s_0), \xi(s_0) + \tau_j \nu)} > 0$$

implies that Γ_j enters D as x_0 increases, and hence, by strict hyperbolicity, $x_0 > x_0(s_0)$ on Γ_j until Γ_j again intersects ∂D.

To verify the remark made earlier that if the matrix $\left(b_i\left(x, \frac{\partial \psi_j}{\partial x}\right)\right)$, $i = 1, \ldots, l, j = 1, \ldots, r = k_0 + (m - k)/2$, has a nontrivial nullspace and rank l, the Cauchy problem is ill-posed, one need only carry out the construction of this section with $a_0, \ldots, a_N$ identically zero and $(a_0^1(x(s_0)), \ldots, a_0^r(x(s_0))$ equal to the nontrivial null vector. Taking R sufficiently large the estimates given here will show that no estimate of the form

$$\int_{D \times R \cap |x - x(s_0)| < \delta} |u(x)|^2 \, dx \leqslant c \left(\|Pu\|_s + \sum_{i=1}^{l} \||B_i u\||_{s'} \right.$$

$$\left. + \sum_{i=1}^{m-1} \left\| \frac{\partial^i u}{\partial t^i}(x',0) \right\|_{s''} \right)$$

can hold for $\delta > 0$. We omit the details.

3. THE NOTION OF WAVE FRONT SET

Wave front sets are tied to Fourier transforms, and we will begin by recalling a few facts about the Fourier transform in R^n. Given $f \in L^1(R^n)$ we define the Fourier transform

$$\hat{f}(\xi) = (2\pi)^{-n/2} \int_{R^n} e^{-ix \cdot \xi} f(x) \, dx. \tag{3.1}$$

If $f \in C_0^\infty(R^n)$, the identity

$$(1 + |\xi|^2)^M \hat{f}(\xi) = (2\pi)^{-n/2} \int_{R^n} e^{ix \cdot \xi} (1 - \Delta)^M f(x) \, dx$$

shows that for each N, there is a C_N such that

$$|\hat{f}(\xi)| \leqslant C_N (1 + |\xi|)^{-N}. \tag{3.2}$$

If both f and $\hat{f}$ are in $L^1(R^n)$, then one has the Fourier inversion formula (see [23], Theorem 9.11)

$$f(x) = (2\pi)^{-n/2} \int_{R^n} e^{ix \cdot \xi} \hat{f}(\xi) \, d\xi \qquad \text{a.e.} \tag{3.3}$$

In particular, if $\hat{f}$ satisfies the estimates (3.2) one may conclude from (3.3) that $f(x) \in C^\infty(R^n)$. Bearing in mind that one can build nonzero functions in $C_0^\infty(R^n)$ with support in any prescribed ball, one sees that smoothness near x_0 can be characterized as follows:

A function $f \in L^1_{\text{loc}}(R^n)$ is equivalent to a C^∞ function in a neighborhood of x_0 if and only if there is a nonnegative $\rho \in C_0^\infty(R^n)$ with $\rho(x_0) = 1$ such that $\widehat{\rho f}(\xi)$ satisfies the estimates (3.2).

Actually this characterization holds for all f in the distribution space $\mathcal{D}'(R^n)$ of continuous linear functionals on $C_0^\infty(R^n)$. Given $f \in \mathcal{D}'(R^n)$ one defines

$$\widehat{\rho f}(\xi) = f\left(\rho e^{-ix\cdot\xi}\right)$$

and says f is equivalent to a C^∞ function g on $\mathcal{O} \subset R^n$, if for all $\phi \in C_0^\infty(\mathcal{O})$

$$f(\phi) = \int_{R^n} g\phi \, dx.$$

With these definitions the characterization can be derived using the Plancherel theorem and basic properties of the topology on $C_0^\infty(R^n)$, (see [8]).

Given $f \in \mathcal{D}'(R^n)$ one defines the "singular support" of f as the complement of the union of the open sets on which f is equivalent to a C^∞ function. In [11] Hörmander introduced the wave front set of f, denoted by $\mathrm{WF}(f)$, as a refinement of the singular support. Heuristically the wave front set tells you not only where f is not smooth but also in which directions it is not smooth. The precise definition can be stated as follows:

We say an open set N in R^n is conic if $\xi \in N \Rightarrow t\xi \in N$ for $t > 0$. Given $(x_0, \xi_0) \in R^n \times R^n/0$ we say $(x_0, \xi_0) \notin \mathrm{WF}(f)$ if there is a $\rho \in C_0^\infty(R^n)$ with $\rho(x_0) = 1$ and a conic neighborhood N of ξ_0 such that $\widehat{\rho f}(\xi)$ satisfies (3.2) for $\xi \in N$. The characterization of smoothness given earlier shows that $\mathrm{WF}(f)$ really is a refinement of singular support:

$$x \in \text{singular support } f \Leftrightarrow (x, \xi) \in \mathrm{WF}(f) \text{ for some } \xi \in R^n/0.$$

To get an idea of the significance of WF(f), consider $f \in L^1_{\text{loc}}(R^n)$ given by

$$f = \begin{cases} 1 & x_1 > 0, \\ 0 & x_1 < 0. \end{cases}$$

Letting $\xi = (\xi_1, \xi')$, $x = (x_1, x')$, $\Delta = \dfrac{\partial^2}{\partial x_1^2} + \Delta'$, we have

$$\left(1 + |\xi'|^2\right)^N \widehat{\rho f}(\xi) = (2\pi)^{-n/2} \int_0^\infty dx_1 \int_{R^{n-1}} e^{ix \cdot \xi} (1 - \Delta')^N \rho f \, dx'$$

and

$$\widehat{\rho f}(\xi_1, 0) = \frac{(2\pi)^{-n/2}}{i\xi_1} \int_{R^{n-1}} \rho(0, x') \, dx' + \frac{1}{i\xi_1} \int_0^\infty dx_1 \int_{R^{n-1}} e^{ix_1 \xi_1} \frac{\partial \rho}{\partial x_1} \, dx.$$

These identities show that WF(f) = $\{(x, \xi) \in R^n \times R^n/0 : x' = \xi' = 0\}$. Thus, if we think of the jump in f along $x_1 = 0$ as a "wave front" the wave front set consists of the pairs (x, ξ) where x is on the wave front and ξ is normal to the wave front. More generally, if $f \in L^1_{\text{loc}}(R^n)$ is smooth up to a smooth surface $\Sigma = \{x | \phi(x) = 0\}$ from both sides but may jump across Σ,

$$\text{WF}(f) \subset \left\{ (x, \xi) : x \in \Sigma, \xi = t \frac{\partial \phi}{\partial x}(x), t \in R/0 \right\}.$$

For more general singularities, the wave front set is not so predictable. The reader may verify that for the functions f_1 and f_2 on R^2, with values in the quadrants as indicated in Figure 1, the wave

<table>
<tr><td>x_2</td><td>x_2</td></tr>
<tr><td>$\dfrac{1 \mid 0}{2 \mid 1} x_1$</td><td>$\dfrac{0 \mid 0}{2 \mid 0} x_1$</td></tr>
<tr><td>values of f_1</td><td>values of f_2</td></tr>
</table>

FIG. 1

front sets are as follows:

$$WF(f_1) = \{(x, \xi): x_1 = 0 \text{ and } \xi_2 = 0, \text{ or } x_2 = 0 \text{ and } \xi_1 = 0\}$$

$$WF(f_2) = \{(x, \xi): x_1 < 0, x_2 = 0 \text{ and } \xi_1 = 0, \text{ or } x_2 < 0, x_1 = 0$$

$$\text{and } \xi_2 = 0, \text{ or } x_1 = x_2 = 0 \text{ and } \xi \in R^2/0\}.$$

If we apply a linear differential operator with smooth coefficients to f, we do not increase its wave front set. To see this, consider first the effect of multiplying f by $\psi \in C^\infty(R^n)$. If $(x_0, \xi_0) \notin WF(f)$ we have $\rho \in C_0^\infty(R^n)$ and a conic neighborhood N of ξ_0 such that $\widehat{\rho f}(\xi)$ satisfies (3.2) for $\xi \in N$. Choosing $\rho' \in C_0^\infty(R^n)$ so that $\rho' = 1$ on support ρ and setting $\psi' = \rho' \psi$, we have

$$\widehat{\rho(\psi f)}(\xi) = \widehat{\psi' \rho f}(\xi) = (2\pi)^{-n/2} \int \hat{\psi}'(\eta) \widehat{\rho f}(\xi - \eta) \, d\eta,$$

where in the last equality we used the convolution formula ([23], Theorem 9.2). If $f \in L^1_{\text{loc}}$, then $|\widehat{\rho f}(\xi)| \leqslant C$ for all ξ. More generally if $f \in \mathcal{D}'(R^n)$, there is an N_0 such that $|\widehat{\rho f}(\xi)| \leqslant C(1 + |\xi|)^{N_0}$. Thus, since $|\hat{\psi}'(\eta)|$ satisfies (3.2), we have

$$\left| \int \hat{\psi}'(\eta) \widehat{\rho f}(\xi - \eta) \, d\eta \right| \leqslant \left| \int_{\delta|\xi| > |\eta|} \hat{\psi}'(\eta) \widehat{\rho f}(\xi - \eta) \, d\eta \right|$$

$$+ \left| \int_{\delta|\xi| < |\eta|} \hat{\psi}'(\eta) \widehat{\rho f}(\xi - \eta) \, d\eta \right|$$

$$\leqslant \sup_{\delta|\xi| > |\eta|} |\widehat{\rho f}(\xi - \eta)| \int_{R^n} |\hat{\psi}'(\eta)| \, d\eta$$

$$+ C \int_{\delta|\xi| < |\eta|} (1 + |\eta|)^{-R} (1 + |\xi - \eta|)^{N_0} \, d\eta.$$

One checks that for $R = M + N_0 + n$, $M > 0$, the second integral is bounded by a multiple of $(1 + |\xi|)^{-M}$. Since N is conic there is a

$\beta > 0$, such that

$$\left| \frac{\xi - \eta}{|\xi - \eta|} - \frac{\xi_0}{|\xi_0|} \right| < \beta \Rightarrow \xi - \eta \in N.$$

Hence, choosing $\delta < \frac{1}{2}$ small enough that $\delta|\xi| > |\eta|$ implies

$$\left| \frac{\xi - \eta}{|\xi - \eta|} - \frac{\xi}{|\xi|} \right| < \beta/2,$$

we conclude that if $\left| \dfrac{\xi}{|\xi|} - \dfrac{\xi_0}{|\xi_0|} \right| < \beta/2$ then

$$\sup_{\delta|\xi| > |\eta|} |\widehat{\rho f}(\xi - \eta)| \leqslant C_M (1 + |\xi|)^{-M}.$$

Since $\{\xi : |\xi/|\xi| - \xi/|\xi_0||< \beta/2\}$ is a conic neighborhood of ξ_0, we have shown $(x_0, \xi_0) \notin \mathrm{WF}(\psi f)$. Given this special case and the identity

$$\widehat{\frac{\partial \rho f}{\partial x_j}}(\xi) = i\xi_j \widehat{\rho f}(\xi),$$

the proof that $\mathrm{WF}(Pf) \subset \mathrm{WF}(f)$ for arbitrary linear differential operator P is easy and we leave it as an exercise.

The main point of the *propagation* of singularities is that, while the inclusion $\mathrm{WF}(Pf) \subset \mathrm{WF}(f)$ is actually an equality when P is an elliptic operator, it is often a strict inclusion when P is hyperbolic.

4. PROPAGATION OF SINGULARITIES

We are going to use the gaussian beams constructed in Section 2 to prove a propagation of singularities theorem for the following mixed problem:

$$\begin{aligned}
Pu &= f \text{ in } R_+ \times \Omega, \\
B_i u &= g_i \text{ on } R_+ \times \partial\Omega, \quad i = 1, \ldots, l, \\
\frac{\partial^i u}{\partial t^i} &= h_i \text{ on } \{0\} \times \Omega, \quad i = 0, \ldots, m-1.
\end{aligned} \qquad (4.0)$$

The proofs of the theorems on the propagation of singularities in the references given earlier rely on pseudodifferential operator techniques, and hence require a good bit of Fourier analysis. The last vestige of Fourier analysis in the approach taken here is in the following lemma.

LEMMA (4.1). *Assume* $\dfrac{\partial \psi}{\partial x} \neq 0$ *and* $(\partial \psi / \partial x)(x_0) = \xi_0$ *and* $(x_0, -\xi_0) \notin \mathrm{WF}(u)$, $u \in L^2_{\mathrm{loc}}$. *Assume* $\phi \in C_0^\infty(R^n)$ *and* $\mathrm{Im}\, \psi \geq c^2 |x - x_0|^2$ *on the support of* ϕ. *Then there are constants* C_N *such that*

$$\left| \int e^{ik\psi} \phi u\, dx \right| \leq C_N k^{-N} \tag{4.2}$$

for $k > 0$ *and* $N \in \mathbb{Z}_+$.

Proof. By the definition of the wave front set there is a $\rho \in C_0^\infty(R^n)$ with $\rho = 1$ on a neighborhood of x_0 such that

$$|\widehat{\rho u}(\xi)| \leq C_N (1 + |\xi|)^{-N} \tag{4.3}$$

for all ξ satisfying $|\xi/|\xi| + \xi_0/|\xi_0|| < \delta$ for some $\delta > 0$. By continuity there is a $\delta' > 0$ such that for $|x - x_0| < \delta'$,

$$\left| k \frac{\partial \psi}{\partial x}(x) + \xi \right|^2 > c(k^2 + |\xi|^2),\, c > 0$$

for all ξ satisfying $|\xi/|\xi| + \xi_0/|\xi_0|| > \delta/2$. Choose $b(\xi)$ homogeneous of degree zero for $|\xi| > 1$, vanishing when $|\xi/|\xi| + \xi_0/|\xi_0|| \leq \delta/2$ and when $|\xi| < \frac{1}{2}$, and equal to one when $|\xi| > 1$ and $|\xi/|\xi| + \xi_0/|\xi_0|| \geq \delta$. Next define Bw by

$$(Bw)(x) = (2\pi)^{-n/2} \int e^{ix \cdot \xi} b(\xi) \hat{w}(\xi)\, d\xi.$$

Choosing $\rho_0 \in C_0^\infty(|x - x_0| < \delta')$ such that $\rho_0 = 1$ on a neighborhood of x_0, one has

$$\int e^{ik\psi}\phi u\, dx = \int e^{ik\psi}(1-\rho_0\rho)\phi u\, dx + \int e^{ik\psi}\phi\rho_0(\rho u - B(\rho u))\, dx$$

$$+ \int e^{ik\psi}\phi\rho_0 B(\rho u)\, dx$$

$$\triangleq I_1 + I_2 + I_3.$$

The integral I_1 automatically satisfies the estimates (4.2), since $\operatorname{Im}\psi > 0$ on the support of $(1-\rho_0\rho)\phi$. By (4.3) the function $f = \rho u - B(\rho u)$ belongs to $C^\infty(R^n)$. Hence, defining a differential operator L by

$$Lw = \left|\frac{\partial\psi}{\partial x}\right|^{-2}\frac{\partial\psi}{\partial x}\cdot\frac{\partial w}{\partial x},$$

one has

$$I_2 = \frac{1}{(ik)^N}\int (L^N e^{ik\psi})\phi\rho_0 f\, dx = \frac{1}{(ik)^N}\int e^{ik\psi}(L^t)^N(\phi\rho_0 f)\, dx,$$

which shows that I_2 satisfies (4.2). For I_3 write (formally—note that the integral in ξ may diverge)

$$I_3 = (2\pi)^{-n/2}\int dx\int e^{i(\xi\cdot x + k\psi)}\rho_0(x)\phi(x)b(\xi)(\widehat{\rho u})(\xi)\, d\xi,$$

and L be the operator

$$Lw = \left|k\frac{\partial\psi}{\partial x}+\xi\right|^{-2}\left(k\frac{\overline{\partial\psi}}{\partial x}+\xi\right)\cdot\frac{\partial w}{\partial x}.$$

Then, again formally,

$$I_3 = i^M\int dx\int e^{-i(\xi\cdot x + k\psi)}(L^t)^M(\rho_0\phi)b(\xi)(\widehat{\rho u})(\xi)\, d\xi, \quad (4.4)$$

which shows that (4.2) holds for I_3. To justify the formal derivation of (4.4) given here, one should replace $(\widehat{\rho u})(\xi)$ by its restriction to $|\xi| < R$. Then the formal manipulations are valid. Since

$$\left|(L^t)^M(\rho_0 \phi)\right| \leq C(k^2 + |\xi|^2)^{-M/2}$$

on the support of $b(\xi)$, one can then let R tend to infinity and recover (4.4). This completes the proof.

Remark (4.5). Note that if ψ depends smoothly on parameters, i.e.,

$$\psi = \psi(x; y, \eta) \text{ and } \frac{\partial \psi}{\partial x}(x_0; y_0, \eta_0) = \xi_0,$$

then the constants in (4.2) are uniform for (y, η) in $|y - y_0| < \delta_0$, $|\eta - \eta_0| < \delta_0$. As in the preceding argument, given $\delta > 0$, there is a $\delta' > 0$ such that

$$|y - y_0| < \delta', |\eta - \eta_0| < \delta', |x - x_0| < \delta',$$

and $|\xi/|\xi| + \xi_0/|\xi_0|| > \delta/2$ imply that

$$\left|k\frac{\partial \psi}{\partial x}(x; y, \eta) + \xi\right| > c(k^2 + |\xi|^2)^{1/2}.$$

Likewise, assuming $\operatorname{Im}\psi(x; y, \eta) \geq c|x - x(y, \eta)|^2$ on the support of ϕ for

$$|y - y_0| < \delta'', |\eta - \eta_0| < \delta'', x(y_0, \eta_0) = x_0,$$

and $x(y, \eta)$ is smooth, there is a $\delta_0 < \min(\delta', \delta'')$ such that $\operatorname{Im}\psi(x; y, \eta) > a > 0$, when

$$|y - y_0| < \delta_0, |\eta - \eta_0| < \delta_0 \text{ and } |x - x_0| > \delta'.$$

Since these two estimates were all that was used in the proof of Lemma (4.1), the generalization is immediate.

Returning now to the mixed problem (4.0), we consider both (4.0) and an adjoint problem. The adjoint problem is defined as follows: the adjoint operator P^* is the (unique) differential operator such that

$$\int_{R^{n+1}} \bar{v}Pu = \int_{R^{n+1}} \overline{P^* v}\, u$$

for all $u, v \in C_0^\infty(R^{n+1})$. Since the principal symbol $p_m(x, \xi)$ of P is real-valued, the principal symbol of P^* is equal to p_m. For our purposes a set of boundary conditions $B_i^* v = 0$, $i = l+1,\ldots,m$, is adjoint to the boundary conditions $B_i u = 0$, $i = 1,\ldots,l$, if there are linear differential operators C_i, $i = 1,\ldots,m$, such that

$$\int_{R \times \Omega} \left(\bar{v}Pu - \overline{P^* v}\, u \right) dx = \int_{R \times \partial\Omega} \left(\sum_{i=1}^{l} \overline{C_i v}\, B_i u + \sum_{i=l+1}^{m} \overline{B_i^* v}\, C_i u \right)$$

for all $u, v \in C_0^\infty(R^{n+1})$. We always assume that all the differential operators B_i, B_i^*, and C_i have order less than m. In the simple case that $B_i u = \partial u^{i-1}/\partial v^{i-1}$, $i = 1,\ldots,l$, where $\partial/\partial v$ is the normal derivative to $\partial\Omega$, the operators $B_i^* u = \partial^{i-1} u/\partial v^{i-1}$, $i = 1,\ldots,m-l$, define adjoint boundary conditions.

At this point it is convenient to reintroduce the time variable suppressed at the beginning of Section 2. Thus we have variables $(t, x) = (t, x_1,\ldots,x_n)$ where $t = x_0$ in the notation of Sections 2 and 3.

Given $T > 0$ and $(\underline{x}, \underline{\xi})$, $\underline{x}$ in the interior of Ω and $|\underline{\xi}| = 1$, let τ_i, $i = 1,\ldots,m$, be the roots of $p_m(T, \underline{x}, \tau, \underline{\xi}) = 0$ and consider the (null) bicharacteristics emanating from $(T, \underline{x}, \tau_i, \underline{\xi})$, $i = 1,\ldots,m$. We follow these bicharacteristics backwards, i.e., into $t < T$. At points (t', x', τ', ξ') on these curves, where $x' \in \partial\Omega$ and $t' < T$, we continue backwards along the bicharacteristics emanating from (t', x', τ', ξ_i), $i = 1,\ldots,r$. Here $\xi_i = \xi' + s_i v$, where v is the inner unit normal to $\partial\Omega$ at x', and s_i, $i = 1,\ldots,r$, are the real roots of $p_m(t', x', \tau', \xi' + sv) = 0$ satisfying

$$0 > v \cdot \left. \frac{\partial p_m}{\partial \xi} \frac{\partial p_m}{\partial \tau} \right|_{(t', x', \tau', \xi' + s_i v)}.$$

Continuing in this way (stopping at points with $x' \in \partial\Omega$ and no admissible s_i), we arrive at a set of points $(0, y_i, \tau_i, \xi_i)$, $i = 1, \ldots, M$, which are the data at time $t = 0$ for the bicharacteristics in the "shower" produced by tracing backwards from $(T, \underline{x}, \underline{\xi})$. We assume that

   (i)   all bicharacteristics in the shower never graze and either stop or reach $t = 0$ after a finite number of reflections,

   (ii) y_i, $i = 1, \ldots, M$, lie in the interior of Ω,

   (iii) the adjoint boundary conditions satisfy the second hypothesis used in the construction of gaussian beams, namely that the matrix $(b^*)_{ij}$ has rank $m - l$. This is the analogue of the hypothesis that the boundary conditions be "perfectly reflecting" which is used in [15].

Then we have

THEOREM (4.6). *Suppose $u \in H^m([0, T] \times \Omega)$ satisfies (4.0). Assume*

   (a) *$(y_i, \xi_i) \notin \mathrm{WF}(h_j)$, $i = 1, \ldots, M$, and $j = 0, \ldots, m - 1$,*

   (b) *for all points (t', x', τ', ξ') on bicharacteristics in the shower with $x' \in \partial\Omega$, the g_j are smooth in a neighborhood of (t', x'), $j = 1, \ldots, l$,*

   (c) *f is supported in the interior of $[0, T] \times \Omega$ and $\mathrm{WF}(f)$ does not intersect the shower.*

Then $(x, \xi) \notin \mathrm{WF}\left(\dfrac{\partial^r u}{\partial t^r}(T, \cdot)\right)$ for $r = 0, \ldots, m - 1$ (and hence all r).

Remark (4.7). The assumption that $u \in H^m([0, T] \times \Omega)$ was made so that u would be a classical solution to the boundary value problem (4.0). What the proof really requires is the identity

$$\int_{[0, T] \times \Omega} (\bar{v}f - \overline{P^*v}\,u) = \int_{[0, T] \times \partial\Omega} \sum_{i=1}^{l} (\overline{C_i v}) g_i + \sum_{i=l+1}^{m} \overline{B_i^* v}\, C_i u$$

$$+ \int_{\langle T \rangle \times \Omega} \sum_{i=1}^{m} \frac{\overline{\partial^{i-1} v}}{\partial t^{i-1}} D_i u + \int_{\langle 0 \rangle \times \Omega} \sum_{i=1}^{m} \overline{E_i v}\, h_i$$

$$(4.8)$$

for all $v \in C^\infty(R^{n+1})$ which vanish near the corners $\{T\} \times \partial\Omega$ and $\{0\} \times \partial\Omega$. This follows from (4.0) by integration by parts when $u \in H^m([0, T] \times \Omega)$. To see that the boundary terms at $t = T$ and $t = 0$ have the form shown in (4.8) for appropriate differential operators E_i and D_i, one uses the cases $l = m$ and $l = 0$ of the remark made earlier that

$$B_i^* v = \partial^{i-1} v / \partial \nu^{i-1}, i = 1, \ldots, m - l,$$

are adjoint to

$$B_i u = \partial^{i-1} u / \partial \nu^{i-1}, i = 1, \ldots, l.$$

Actually it is easy to extend the proof of Theorem (4.6) to the case where u, f, g_i and h_i are distributions and (4.8) holds in the distribution sense.

Remark (4.9).   In (b) of Theorem (4.6) we have not made the hypotheses on the g_j in terms of wave front sets. This could be done easily but it would entail defining the wave front set of a distribution on a manifold M—in this case $M = R \times \partial\Omega$. This is done at the outset in Hörmander [11]. The wave front is invariantly defined as a subset of the cotangent bundle $T^*(M)$.

Remark (4.10).   The nongrazing hypothesis used here, namely that the roots of $p_m(t', x', \tau', \xi + s\nu) = 0$ are distinct, is not natural. The natural hypothesis, used in [15], [18], and [27], is that the *real* roots of this equation are distinct. The extraneous hypothesis here arises because one cannot build gaussian beams using the ansatz (2.10) when the complex roots are multiple.

Proof of Theorem (4.6).   Given Lemma (4.1) and the construction of gaussian beams, the proof of this theorem is easy. We choose $\delta > 0$ sufficiently small that the shower obtained by tracing backward from (T, x, ξ) when $|x - \underline{x}| < \delta$ and $|\xi - \underline{\xi}| < \delta$ has the properties assumed for the shower obtained from $(\overline{T}, \underline{x}, \underline{\xi})$. Then, given

N, we construct gaussian beams $\omega(t, x; x', \xi, k)$ satisfying for $k > 0$

(i) $\|P^*\omega\|_0 \leqslant Ck^{-N}$

(ii) $\|B_i^*\omega\|_0 \leqslant Ck^{-N}, \quad i = l+1,\ldots,m$ \qquad (4.11)

(iii) $\left\|\dfrac{\partial^r\omega}{\partial t^r}(T, x)\right\|_0 \leqslant Ck^{-N}, \quad r = 0,\ldots,m-2, \quad$ and

$$\left\|\frac{\partial^{m-1}}{\partial t^{m-1}}(T, x, x', \xi) - k^{m-1}\phi(x)\exp\left(ikx\cdot\xi - \frac{k}{2}|x - x'|^2\right)\right\|_0$$

$$\leqslant Ck^{-N}$$

where $\phi \in C_0^\infty(|x - \underline{x}| < \delta)$ and $\phi(\underline{x}) = 1$. For t close to T, ω has the form

$$\omega = \sum_{j=1}^m e^{ik\psi_j}\left(a_0^j + \cdots + \frac{a_R^j}{k^R}\right),$$

where

$$\left(\frac{\partial\psi_j}{\partial t}(T, x'), \frac{\partial\psi_j}{\partial x}(T, x')\right) = \left(\tau_j(x', \xi), \xi\right)$$

and

$$p_m\left(T, x', \tau_j(x, \xi), \xi\right) = 0, \quad j = 1,\ldots,m.$$

To see that it is possible to choose the coefficients $a_0^j,\ldots,a_R^j$, $j = 1,\ldots,m$, so that (4.11) (iii) holds, one observes that the Taylor series about x' of $a_r^j(T, x)$ is determined by equations of the form

$$\sum_{j=1}^m \left(\tau_j(x, \xi)\right)^i a_r^j(T, x) = g_{ri}, \qquad i = 0,\ldots,m-1, \quad (4.12)$$

where g_{ri} is determined by ϕ, the ψ_j, and the $a_{r'}^j$, $r' < r$. Since by strict hyperbolicity the τ_j's are distinct, the system (4.12) is always uniquely solvable ($A_{ij} = (\tau_j)^i$ is a Vandermonde matrix). Thus we can construct ω satisfying (4.11) (i)–(iii) and the smooth dependence of ω on (x', ξ) implies that for δ sufficiently small the constant C in (4.11) is uniform for $|x' - \underline{x}| < \delta$, $|\xi - \underline{\xi}| < \delta$.

Since $\underline{x}$ and y_i, $i = 1,\ldots,M$, lie in the interior of Ω, ω can be constructed to vanish near the corners $\{T\} \times \partial\Omega$ and $\{0\} \times \partial\Omega$.

Hence (4.8) holds with $v = \omega$. Since the coefficient of $\partial^m/\partial t^m$ in P was one, $D_m u = (-1)^{m-1} u$. Thus by Lemma (4.1) and (4.11) we can rewrite (4.8) as

$$\int_{[0,T] \times \Omega} \overline{\omega} f = (-1)^{m-1} \int_{\Omega} k^{m-1} \phi(x) \exp\left(-ikx \cdot \xi - \frac{k}{2} |x - x'|^2\right)$$

$$\times u(x, T) \, dx + O(k^{-N}) \tag{4.13}$$

where $O(k^{-N})$ is uniform for x' in neighborhood of support ϕ and $|\xi - \underline{\xi}| < \delta$. Some comment needs to be made about the use of Lemma (4.1) to eliminate the terms in (4.8) involving the g_j's. Here one must introduce local coordinates $y_1, \dots, y_{n-1}$ on $\partial\Omega$ and apply the lemma to $g_j(t, x(y))$ with the phase function $\psi = \psi(t, x(y))$. Since $v \cdot (\partial x/\partial y) = 0$ and $(\partial x/\partial y)$ has rank $n - 1$, it follows that $\left(\dfrac{\partial \psi}{\partial t}, \dfrac{\partial \psi}{\partial y}\right) = 0$ only if $\left(\dfrac{\partial \psi}{\partial t}, \dfrac{\partial \psi}{\partial x}\right) = (0, \alpha v)$, but this is ruled out by the nongrazing hypothesis. Since $g_j(t, x(y))$ is smooth near (t', x'), its wave front set is empty and Lemma (4.1) applies.

Obvious modifications in the proof of Lemma (4.1) yield the following: given $t_0 \in (0, T)$ there is an $\varepsilon > 0$ such that for $\rho \in C_0^\infty(|t - t_0| < \varepsilon)$

$$\left| \int_{[0,T] \times \Omega} \rho \overline{\omega} f \right| \leqslant C_N k^{-N}.$$

Combining this with (4.13), we arrive at last at

$$\int_{\Omega} \phi(x) \exp\left(-ikx \cdot \xi - \frac{k}{2} |x - x'|^2\right) u(x, T) \, dx \leqslant C_N k^{-N}$$

where C_N is uniform for x' in a neighborhood $\mathcal{O}$ of support ϕ and $|\xi - \underline{\xi}| < \delta$. Thus

$$\left| \int_{\Omega} \phi(x) e^{-ikx \cdot \xi} \left(k^{n/2} \int_{\mathcal{O}} e^{-k/2 |x - x'|^2} \, dx' \right) u(x, T) \, dx \right| \leqslant C_N k^{-N+n/2}.$$

Finally, letting $C = \int_{R^n} e^{-|y|^2/2} \, dy$, we see that

$$\left| \phi(x) \left(C - k^{n/2} \int_{\Theta} e^{-k/2|x-x'|^2} \, dx' \right) \right| \leqslant C_N k^{-N}$$

uniformly for x in R^n. Thus

$$\left| \int_{\Omega} \phi(x) e^{-ikx \cdot \xi} u(x, T) \, dx \right| \leqslant C_N k^{-N}$$

for each N, uniformly for $|\xi - \underline{\xi}| < \delta$, and we have shown that $(\underline{x}, \underline{\xi}) \notin \mathrm{WF}(u(x, T))$.

To show that $(\underline{x}, \underline{\xi}) \notin \mathrm{WF}\left(\dfrac{\partial u}{\partial t}(x, T) \right)$ one returns to (4.8) and notes that

$$D_{m-1} u = (-1)^{m-2} \frac{\partial u}{\partial t} + F(u)$$

where F is a differential operator in the x-variables alone. Thus $(\underline{x}, \underline{\xi}) \notin \mathrm{WF}(F(u))$. Choosing a new gaussian beam ω so that

$$\left\| \frac{\partial^r \omega}{\partial t^r}(T, x) \right\|_0 \leqslant C_N k^{-N}, \qquad 0 \leqslant r \leqslant m-1, \qquad r \neq m-2,$$

$$\left\| \frac{\partial^{m-2} \omega}{\partial t^{m-2}}(T, x) - k^{m-2} \phi(x) \exp\left(ikx \cdot \xi - \frac{k}{2}|x - x'|^2 \right) \right\|_0 \leqslant C_N k^{-N}$$

one can use the argument just given to show

$$\left| \int_{\Omega} \phi(x) e^{-ikx \cdot \xi} (D_{m-1}(u)(x, T) \right| \leqslant C_N k^{-N}.$$

Since ϕ can be chosen at the outset so that

$$\left| \int_{\Omega} \phi e^{-ikx \cdot \xi} F(u)(x, T) \, dx \right| \leqslant C_N k^{-N}$$

uniformly for $|\xi - \underline{\xi}| < \delta$, one may conclude

$$(\underline{x}, \underline{\xi}) \notin \mathrm{WF}\left(\frac{\partial u}{\partial t}(x, T)\right).$$

The same observations apply to $\dfrac{\partial^r u}{\partial t^r}(x, T)$, $r = 2, \ldots, m - 1$, and they complete the proof.

5. AN EXAMPLE AND AN APPLICATION

Gaussian beams for the standard wave equation

$$0 = \frac{\partial^2 u}{\partial t^2} - \left(\frac{\partial^2 u}{\partial x_1^2} + \frac{\partial^2 u}{\partial x_2^2}\right) \equiv \Box u$$

are easy to construct and they may be helpful in understanding the construction of Section 2. The initial value problem for this equation is well posed in the "energy norm" given by

$$\left\|\left\{u(t, \cdot), \frac{\partial u}{\partial t}(t, \cdot)\right\}\right\|_E^2 = \int_{R^2} \left|\frac{\partial u}{\partial t}(t, x)\right|^2 + \left|\frac{\partial u}{\partial x}(t, x)\right|^2 dx,$$

and, if $\Box u = f$, then for $0 \leqslant t \leqslant T$

$$\left\|\left\{u(t, \cdot), \frac{\partial u}{\partial t}(t, \cdot)\right\}\right\|_E \leqslant \left\|\left\{u(0, \cdot), \frac{\partial u}{\partial t}(0, \cdot)\right\}\right\|_E + T \sup_{[0, T]} \|f(t, \cdot)\|_0.$$

$$(5.0)$$

In the notation of Section 2 the bicharacteristic equations here are

$$\dot{x}_0 = 2\xi_0, \quad \dot{x}_i = -2\xi_i, \quad i = 1, 2, \quad \text{and} \quad \dot{\xi} = 0.$$

We consider the null bicharacteristic

$$(x(s), \xi(s)) = \left(s, 0, s, \frac{-1}{2}, 0, \frac{1}{2}\right).$$

Making the ansatz

$$u = e^{ik\psi}a_1,$$

with $\operatorname{Im}\psi \geqslant 0$, and proceeding as in Section 2 we get

$$\frac{\partial \psi}{\partial x}(s,0,s) = \left(\frac{-1}{2},0,\frac{1}{2}\right) \quad \text{and} \quad \frac{dM}{ds} + 2MAM = 0,$$

where

$$(M(s))_{ij} = \frac{\partial^2 \psi}{\partial x_i \partial x_j}(s,0,s) \text{ and } A = \begin{pmatrix} -1 & 0 & 0 \\ 0 & 1 & 0 \\ 0 & 0 & 1 \end{pmatrix}.$$

The equation (2.6) becomes

$$\dot{y} = 2A\eta \quad \dot{\eta} = 0.$$

Hence we have

$$M(s) = M(0)(I + 2sAM(0))^{-1},$$

and we must choose $M(0)$ so that

$$M(0)\begin{pmatrix} 1 \\ 0 \\ 1 \end{pmatrix} = \begin{pmatrix} 0 \\ 0 \\ 0 \end{pmatrix}$$

and $\operatorname{Im} M(0)$ is positive definite on the span of $(0,1,0)$ and $(0,0,1)$. A simple admissible choice is

$$M(0) = \begin{pmatrix} bi & 0 & -bi \\ 0 & ai & 0 \\ -bi & 0 & bi \end{pmatrix}$$

where a and b are positive constants. Then a ψ with the proper 1st and 2nd partials on $(s,0,s)$ is

$$\psi = \frac{x_2 - x_0}{2} + \frac{a^2 x_0 x_1^2}{1 + 4a^2 x_0^2} + i\left(\left(\frac{a}{1 + 4a^2 x_0^2}\right)\frac{x_1^2}{2} + b\frac{(x_2 - x_0)^2}{2}\right).$$

The equation $c_{-m+1}(x) = 0$ on $x(s) = (s, 0, s)$ from (2.7) reduces to

$$\frac{da_1}{ds} - (\Box \psi) a_1 = 0.$$

On $x(s) = (s, 0, s)$, $\Box \psi = -ia(1 + 2ais)^{-1}$. Thus taking $a_1(0) = 1$, we see that

$$a_1(s) = (1 + 2ais)^{-1/2},$$

where the branch of the square root is chosen so that $a_1(0) = 1$. A choice of $a_1(x_0, x_1, x_2)$ with the proper values on $x(s) = (s, 0, s)$ is

$$a_1 = (1 + 2aix_0)^{-1/2}.$$

As one might guess from the simple form of a_1 there is more structure in the equation $c_{-m+1}(x(s)) = 0$ than we have described here. See [21] for further discussion. At any rate one can check directly that with the choices of ψ and a_1 made here $p_2\left(x, \frac{\partial \psi}{\partial x}(x)\right)$ vanishes to 3rd order on Γ and $c_{-1}(x)$ vanishes to first order on Γ so that Lemma (2.8) implies

$$\|\Box(e^{ik\psi} a_1)\|_0 \leqslant C, \qquad \forall k > 0.$$

Since the energy norm of the initial data

$$\left(e^{ik\psi(0, x_1, x_2)} a_1(0, x_1, x_2), \frac{\partial}{\partial x_0}\left(e^{ik\psi(x_0, x_1, x_2)} a_1(x_0, x_1, x_2)|_{x_0 = 0}\right)\right)$$

is bounded below by a constant multiple of $k^{1/2}$, it follows from (5.0) that the energy of the difference between the constructed and the true solution with the same initial data is small relative to their total energy.

Readers familiar with the more standard geometric optics approximations with real phase functions will note one feature of the approximation here. The level sets $k \operatorname{Im} \psi = 1$ which describe the region where the beam has significant strength are ellipses with axes

$$\left(\frac{2}{kb}\right)^{1/2} \quad \text{and} \quad \left(\frac{2 + 8a^2 x_0^2}{ka}\right)^{1/2}.$$

Thus the beam spreads out in the x_1-direction as $x_0 \to \pm \infty$. A beam that spread this way and had a real phase function would necessarily develop a caustic at some point and after crossing the caustic the phase would appear to have jumped back by $\pi/2$. In the approximate solution here, the factor $a_1(x(s))$ is constantly retarding the phase on $x(s)$ and the total retardation as one goes from $x_0 = -\infty$ to $x_0 = \infty$ is $\pi/2$.

The results we have discussed in Section 4 can be applied rather easily to problems of boundary control for the wave equation as discussed by Russell [24] and Seidman [26]. The problem of boundary control is the following.

Given a domain D in R^n with piecewise C^∞ boundary, one chooses an open subset $\Gamma \subset \partial D$ which is smooth and considers the mixed problem

$$\Box u = 0 \quad \text{in } D \times [0, T],$$

$$u = 0 \quad \text{on } \Gamma \times [0, T], \tag{5.1}$$

$$\left(u(0, x), \frac{\partial u}{\partial t}(0, x)\right) = (f, g).$$

One says the problem is exactly controllable in time T, if for each pair (f, g) of finite energy it is possible to prescribe the values of u on $(\partial D - \Gamma) \times [0, T]$ so that

$$\left(u(T, x), \frac{\partial u}{\partial t}(T, x)\right) = (0, 0).$$

Naturally for this problem to make sense one must specify the set of admissible boundary values for u and the sense in which (5.1) is to hold. Nonetheless the only property that we require of a solution of (5.1) is the following:

Given a solution of (5.1) which has been controlled to zero at $t = T$, there is a distribution $l \in \mathcal{D}'(\Gamma \times R)$ such that for all

$$\omega(t, x) \in C^\infty(R \times R^n)$$

with $\{$support $\omega\} \cap \partial D \times [0, T] \subset \Gamma \times (0, T)$,

$$\int_{D \times [0, T]} \Box \omega u = l(\omega) + \int_D \left(\frac{\partial u}{\partial t}(0, x) \omega(0, x) - u(0, x) \frac{\partial \omega}{\partial t}(0, x) \right) dx.$$

$$(5.2)$$

With this condition one has the following theorem.

THEOREM. *Suppose there is a nongrazing reflected null bicharacteristic for $\Box$ in $D \times R$ which intersects $\partial D \times [0, T]$ only in $\Gamma \times [0, T]$. Then there are initial data (f, g) in $C_0^r(D)$ for each $r < \infty$ such that no solution of (5.1) satisfying (5.2) exists. Thus (5.1) is not exactly controllable to zero in time T.*

Proof. In the notation of Section 2 and Section 4 the boundary condition is defined by B_1, where $B_1 u = u$ and we may assume the adjoint boundary condition is defined by B_1^*, where $B_1^* u = u$. Since $p_2(x, t, \xi, \tau) = |\xi|^2 - \tau^2$, the nongrazing condition reduces to $\nu(x) \cdot \xi \neq 0$. Assuming $(x, t) \in \Gamma \times (0, T)$ and $p(x, t, \xi, \tau) = 0$, the roots of $p(t, x, \xi + s\nu, \tau) = 0$ are $s = 0$ and $s = -2\nu \cdot \xi$. Thus in this case a nongrazing bicharacteristic strikes the boundary with data (x, t, ξ, τ), $\nu \cdot \xi < 0$, and is reflected off with data $(x, t, \xi - 2(\nu \cdot \xi)\nu, \tau)$. Thus the "shower" of Theorem (4.6) reduces to a pair of broken straight line paths, satisfying the familiar "angle of incidence equals angle of reflection" law at the reflection points on $\partial D \times R$. Moreover, since the matrix $(b^*)_{ij}$ of Section 2 reduces to the 1×1 identity matrix here, the nondegeneracy condition of Section 2 is satisfied.

With the preliminaries above out of the way, the idea of the proof is as follows. Given that the hypothesized bicharacteristic has data $(x_0, 0, \xi_0, \tau_0)$, we are going to choose

$$\left(u(x,0), \frac{\partial u}{\partial t}(x,0) \right) = (0, h),$$

where $h \in C^r$ and $(x_0, \xi_0) \in \mathrm{WF}(h)$. This is easy to do; one can choose

$$h = \left(((x - x_0) \cdot \xi_0)_+ \right)^{r+1} \rho(x),$$

where ρ is a cutoff function with $\rho(x_0) = 1$ and

$$(f(x))_+ = \begin{cases} f(x) & \text{if} \quad f(x) > 0 \\ 0 & \text{if} \quad f(x) \leqslant 0. \end{cases}$$

Then, assuming that these initial data can be controlled to zero, we will derive a contradiction from the fact that

$$\left(u(x,T), \frac{\partial u}{\partial t}(x,T) \right) = (0,0)$$

has empty wave front set. One would like to use Theorem (4.6) with the roles of $t = 0$ and $t = T$ reversed for this proof. Unfortunately in trying to do this one encounters the difficulty that there are two null bicharacteristics with data $(x_0, 0, \xi_0, \tau)$ and there is no assurance that the one with data $(x_0, 0, \xi_0, -\tau_0)$ does not intersect $(\partial D - \Gamma) \times [0, T]$. Hence we will not use Theorem (4.6). Instead we will go back to Section 2 and build a nice gaussian beam. The results of Section 2 and our hypothesis on the nature of the null bicharacteristic with data $(x_0, 0, \xi_0, \tau_0)$ imply that we can build a gaussian beam $\omega(x, t, k)$ with the following properties:

(i)   For all k $\langle$support $\omega \rangle \cap \partial D \times [0, T]$ is contained in a fixed compact subset K of $\Gamma \times (0, T)$;

(ii) $\| \Box \omega \|_0 \leqslant C_N k^{-N}$ and $\| |\omega| \|_s \leqslant C_{N,s} k^{-N}$;   (5.3)

(iii) $\omega(x, 0, k) = e^{ikx \cdot \xi_0} e^{-k|x - x_0|^2} \rho_1(x)$ where ρ_1 is a cutoff function with $\rho_1(x_0) = 1$.

The ω with these properties is of the form

$$\omega = e^{ik\psi}\left(a_1 + \frac{a_2}{k} + \cdots + \frac{a_{M+1}}{k^M}\right)$$

for t small, where

$$\left(\frac{\partial\psi}{\partial t}(x_0), \frac{\partial\psi}{\partial x}(x_0)\right) = (\tau_0, \xi_0).$$

Assuming that u is the solution of (5.1) with the initial data above controlled to zero, we may put $\omega(x, t, k)$ into (5.2). Since the distribution l is locally of finite order, we can assume $l \upharpoonright C_0^\infty(K)$ is in H^{-s}. Then (5.2) and (5.3) yield

$$\int_D e^{ikx \cdot \xi_0} e^{-k|x-x_0|^2} \rho_1 h \, dx \leq C_N k^{-N}.$$

The reader can show (by integrating by parts) that this is false for $N > r + 1 + n/2$. This contradiction proves the theorem.

REFERENCES

1. J. A. Arnaud, "Hamiltonian theory of beam mode propagation," *Progress in Optics XI*, ed. E. Wolf, North Holland, 1973, pp. 249–304.
2. ———, *Beam and Fiber Optics*, Academic Press, New York, 1976.
3. K. G. Andersson and R. Melrose, "Propagation of singularities along gliding rays," *Séminaire Goulaouic-Schwartz* 1976–77, Oct. 1976.
4. ———, "The propagation of singularities along gliding rays," *Invent. Math.*, **41** (1977), 197–232.
5. R. Courant and P. D. Lax, "The propagation of discontinuities in wave motion," *Proc. Nat. Acad. Sci. U.S.*, **42** (1956), 872–876.
6. J. J. Duistermaat and L. Hörmander, "Fourier integral operators II," *Acta Math.*, **128** (1972), 183–269.
7. G. Eskin, "Parametrix and propagation of singularities for the interior mixed hyperbolic problem," *J. Analyse Math.*, **32** (1977), 17–62.
8. L. Hörmander, *Linear Partial Differential Operators*, Springer-Verlag, 1964.
9. ———, "The spectral function of an elliptic operator," *Acta Math.*, **121** (1968), 193–218.
10. ———, "Linear differential operators," *Actes Congrès Intern. Math.*, **1** (1970), 121–133.

11. ———, "Fourier integral operators I," *Acta Math.*, **127** (1971), 79–183.
12. ———, "On the existence and regularity of solutions of linear pseudo-differential equations," *Enseign. Math.*, **17** (1971), 99–163.
13. P. D. Lax, "Asymptotic solutions of oscillatory initial value problems," *Duke Math. J.*, **24** (1957), 627–646.
14. D. Ludwig, "Exact and asymptotic solutions of the Cauchy problem," *Comm. Pure Appl. Math.*, **13** (1960), 473–508.
15. A. Majda and S. Osher, "Reflections of singularities at the boundary," *Comm. Pure Appl. Math.*, **28** (1975), 479–499.
16. A. Melin and J. Sjöstrand, "Fourier integral operator with complex-valued phase functions," *Fourier Integral Operators and Partial Differential Equations*, Springer Lecture Notes 459.
17. R. B. Melrose, "Local Fourier-Airy operators," and "Microlocal parametrices for diffractive boundary value problems," *Duke Math. J.*, **42** (1975), 583–604 and 605–632.
18. L. Nirenberg, *Lectures on Linear Partial Differential Equations*, Regional Conference Series #17, A.M.S., Providence, 1973.
19. A. Ja. Povzner and I. V. Suharevskii, "Discontinuities of the Green's function of a mixed problem for the wave equation," *Mat. Sb.*, **51** (93) (1960), 3–20; English transl., *Amer. Math. Soc. Transl.*, (2) **47** (1965), 131–156.
20. J. V. Ralston, "On the propagation of singularities in solutions of symmetric hyperbolic partial differential equations," *Comm. Partial Differential Equations*, **1** (1976), 87–133. "Correction" *Comm. Partial Differential Equations*, **3** (1978), 471–474.
21. ———, "On the construction of quasi-modes associated with stable periodic orbits," *Comm. Math. Phys.*, **51** (1976), 219–242.
22. J. Rauch, "Asymptotic behavior of solutions to hyperbolic partial differential equations with zero speeds," *Comm. Pure Appl. Math.*, **31** (1978), 438–480.
23. W. Rudin, *Real and Complex Analysis*, McGraw-Hill, New York, 1962.
24. D. L. Russell, "Exact boundary controllability theorems for wave and heat processes in star-complemented regions," *Differential Games and Control Theory*, ed. Roxin, Lin, Sternberg, Marcel Dekker, New York, 1974.
25. R. Sakamoto, "Mixed problems for hyperbolic equations, I and II," *J. Math. Kyoto Univ.*, **10** (1970), 349–373 and 403–417.
26. T. I. Seidman, "Exact boundary control for some evolution equations," *SIAM J. Control Optim.*, **16** (1978), 979–999.
27. M. Taylor, "Reflection of singularities of solutions to systems of differential equations," *Comm. Pure Appl. Math.*, **28** (1975), 457–478.
28. ———, "Grazing rays and reflection of singularities of solutions to wave equations," *Comm. Pure Appl. Math.*, **29** (1976), 1–38.
29. G. Uhlmann, "Pseudo-differential operators with involutive double characteristics," *Comm. Partial Differential Equations*, **2** (1977), 713–779. "Parametrices for operators with multiple involutive characteristics," *Comm. Partial Differential Equations*, **4** (1979), 739–767.
30. V. M. Babich and V. S. Buldyrev, *Asymptotic Methods in Short Wave Diffraction Problems* (in Russian), Nauka, Moscow, 1972.

31. H. G. Garnir, ed., *Singularities in Boundary Value Problems*, NATO Advanced Study Institute Series, Reidel, Dordrecht, 1981.
32. R. Melrose and J. Sjöstrand, "Singularities of boundary value problems I", *Comm. Pure Appl. Math.*, **31** (1978), 593–617.
33. ———, "Singularities of boundary value problems II", *Comm. Pure Appl. Math.*, **35** (1982), 129–168.
34. M. Taylor, *Pseudo Differential Operators*, Princeton University Press, Princeton, 1981.

February 1978

REPRESENTATION FORMULAS FOR SOLUTIONS TO $\Delta u - u = 0$ IN R^n

Luis A. Caffarelli* and Walter Littman*

INTRODUCTION

Section 1 of this paper is devoted to a proof of a representation theorem for positive solutions of $\Delta u - u = 0$ in R^n. Section 2 uses Fourier Series methods to obtain the same representation, as well as a weaker representation in the absence of positivity. Section 3 is devoted to Harnack type inequality in R^n of positive solutions to a more general second order elliptic equation.

It should be mentioned that the representation formulas are not new but have been derived by means of Martin boundary considerations and group representation theory, respectively. (See [2] and references there.) Our proofs are essentially elementary with the possible exception of the well-known asymptotic formulas for Bessel functions used in Section 2. The methods of Section 2 have many formal similarities to the methods of papers by C. Müller [5], [6], P. Hartman and C. Wilcox [1], H. Niemeyer [7], dealing with the equation $\Delta u + u = 0$, although the results differ.

---

*Partially supported by NSF Grants.

Sections 1 and 2 may be read independently with the slight exception that the easily derived part (c) of Lemma 1 of Section 1 is used in Section 2.

Results similar to the results of Section 3 were obtained by Landis [4] for more general domains. However, for R^n our results are sharper.

We wish to take this opportunity to thank J. Serrin for bringing this problem to our attention.

1. POSITIVE SOLUTIONS

In this section we intend to prove the following representation theorem:

THEOREM 1. *If u is a positive solution of the equation $\Delta u - u = 0$, there exists a unique measure μ defined on the unit sphere of R^n, such that*

$$u(x) = \int_{|\lambda|=1} e^{\lambda \cdot x} d\mu(\lambda).$$

The idea of the proof is as follows: Assume that u is actually of the form

$$u(x) = \int e^{\lambda \cdot x} f(\lambda) \, dA(\lambda)$$

where $dA(\lambda)$ denotes the usual element of surface area on S^{n-1}. Then, for x going radially to infinity, the function $g_x(\lambda) = e^{\lambda \cdot x - |x|}$ converges to zero everywhere except at $\lambda = x/|x|$ and properly normalized (by $C_x = 1/e^{-|x|}\int e^{\lambda \cdot x} dA(\lambda)$) it becomes a mollifier. This suggests that we should be able to reconstruct $f(\lambda)$ as the limit

$$f(\lambda) = \lim_{\rho \to \infty} \frac{u(\rho\lambda)}{\gamma(\rho)}$$

with

$$\gamma(\rho) = \int_{|\lambda|=1} e^{\rho\lambda \cdot \nu} dA(\lambda) \qquad (\nu \text{ any fixed unit vector}).$$

(The integral does not depend on ν.)

To do that, let us first enumerate some properties of γ, the verification of which we leave as exercises to the reader.

LEMMA 1. (a) *The functions of the form $K\gamma(\rho)$ are the only radial solutions of the problem $\Delta u - u = 0$. (This also follows from equation (6) of Section 2.)*

(b) *For any $\delta < 1$, there exists a positive constant C_δ such that*

$$\gamma(\rho) > C_\delta(e^{\delta\rho}).$$

(c) *For any continuous function $f(\lambda)$ on the unit sphere,*

$$\int_{|\lambda| = 1} \frac{e^{\rho\lambda\cdot\sigma}}{\gamma(\rho)} f(\lambda)\, dA(\lambda) \to f(\sigma) \text{ uniformly as } \rho \to \infty.$$

We also need to mollify u by averaging its rotations.

For that purpose we need to know that the averaged function is again a solution. We first need

LEMMA 2. *Given two continuous functions ϕ and ψ of a single variable; suppose $|x| = |z| = 1$; then the integral*

$$\int_{|y| = 1} \phi(x\cdot y)\psi(y\cdot z)\, dA(y)$$

depends only on the number $x\cdot z$.

Proof. Suppose $x\cdot z = x'\cdot z'$. Then there exists an orthogonal matrix M (i.e., a rotation) such that

$$Mx = x', \qquad Mz = z'.$$

Setting $y' = My$,

$$\int \phi(x'\cdot y)\psi(y\cdot z')\, dA(y) = \int \phi(x\cdot y')\psi(y'\cdot z)\, dA(y')$$

$$= \int \phi(x\cdot y)\psi(y\cdot z)\, dA(y),$$

thus proving the lemma.

METALEMMA. *Let L be a linear operator invariant under rigid motions (that is, $L[\nu(Rx)] = (L\nu)(Rx)$); then if, with $|\sigma| = 1$,*

$$g(\rho\sigma) = \int_{|\lambda|=1} k(\sigma\cdot\lambda) f(\rho\lambda)\, dA(\lambda),$$

$$(Lg)(\rho\sigma) = \int_{|\lambda|=1} k(\sigma\cdot\lambda)(Lf)(\rho\lambda)\, dA(\lambda).$$

To convert this "metalemma" into an honest-to-goodness lemma, the spaces of functions to which ν, $L\nu$, and k belong must be chosen so as to give sense to the operations involved and make them continuous. We shall prove the (meta) lemma for the special case where ν and f are in $C^\infty(R^n)$, k is in $C^\infty(R^1)$ and L is $\Delta - I$. It obviously holds under much more general conditions.

Proof. Let ϕ_ε be a radially symmetric, C_0^∞ mollifier. Then

$$Lg = \lim_{\varepsilon\to 0} L(\phi_\varepsilon * g) = \lim_{\varepsilon\to 0} (L\phi_\varepsilon) * g.$$

It would be enough, therefore, to prove our lemma for $L_\varepsilon(g) = (L\phi_\varepsilon) * g$. Now $(L\phi_\varepsilon)$ is a radially symmetric function and therefore

$$(L\phi_\varepsilon) * g(\rho\sigma) = \int_{\rho'\geqslant 0}\int_{|\sigma'|=1} \mu_\rho(\rho',\sigma\cdot\sigma') g(\rho'\sigma')(\rho')^{n-1}\, dA(\sigma')\, d\rho'$$

where $\mu_\rho(\rho',\sigma\cdot\sigma')$ is simply $(L\phi_\varepsilon)(x - \rho'\sigma')$ written in polar coordinates. Changing the order of integration,

$$(L\phi_\varepsilon) * g(\rho\sigma) = \int_{|\lambda|=1}\int_{\rho'\geqslant 0} f(\rho'\lambda)(\rho')^{n-1}$$

$$\times \int_{|\sigma'|=1} \mu_\rho(\rho',\sigma\cdot\sigma') k(\sigma'\cdot\lambda)\, dA(\sigma')\, d\rho'\, dA(\lambda).$$

But the σ' integral depends on σ and λ only through $\sigma\cdot\lambda$ and hence we can interchange them.

That gives us

$$[(L\phi)_\varepsilon * g](\rho\sigma) = \int k(\sigma \cdot \sigma')[(L\phi_\varepsilon) * f](\rho\sigma') \, dA(\sigma')$$

and the proof is complete.

Finally, as $k(\sigma \cdot \lambda)$ we will use the Poisson kernel $P_\varepsilon(\sigma \cdot \lambda)$ for the unit sphere. Letting $r = e^{-\varepsilon}$, $P_\varepsilon(\sigma \cdot \lambda)$ is the harmonic function inside the unit sphere, which approaches the delta measure at the point λ on the boundary, evaluated at the point with polar coordinates (r, σ). For fixed λ it is a smooth function of σ for $\varepsilon > 0$ approaching the delta function $\delta_\lambda(\sigma)$ on the unit sphere as $\varepsilon \to 0$. We shall make use of the following properties of $P_\varepsilon(\sigma, \lambda)$ for $\varepsilon > 0$:

$$\|P_\varepsilon\|_{L^\infty} \leqslant C(\varepsilon), \qquad \|\partial_\sigma P_\varepsilon\|_{L^\infty} \leqslant C(\varepsilon)$$

where ∂_σ represents differentiation with respect to arc length on the unit sphere $|\sigma| = 1$. We will now consider the functions

$$u_\varepsilon(\rho\sigma) = \int_{|\lambda| = 1} P_\varepsilon(\sigma \cdot \lambda) u(\rho\lambda) \, dA(\lambda).$$

LEMMA 3. *There exists a constant* $C = C(\varepsilon, u)$ *such that*

$$|u_\varepsilon(\rho\sigma)| \leqslant C(\varepsilon, u)\gamma(\rho),$$

$$|\partial_\sigma u_\varepsilon(\rho\sigma)| \leqslant C(\varepsilon, u)\gamma(\rho).$$

Proof.

$$u_1(\rho) = \int u(\rho\lambda) \, dA(\lambda)$$

is a radial solution of $\Delta u - u = 0$ and hence $u_1(\rho) = C(u)\gamma(\rho)$, where $C(u) = u(0)$. The proof now follows from the estimates on P_ε.

From the lemma we conclude for ε fixed that $g_\rho(\sigma) = u_\varepsilon(\rho\sigma)/\gamma(\rho)$ is a bounded family of equicontinuous functions of σ on the unit sphere and therefore we can extract a sequence $\rho_n \to \infty$ such that $g_{\rho_n}(\sigma) \to f_\varepsilon(\sigma)$ uniformly.

LEMMA 4.

$$u_\varepsilon(\rho\sigma) = \int_{|\lambda|=1} f_\varepsilon(\lambda) e^{\rho\lambda\cdot\sigma} \, dA(\lambda).$$

Proof. Let us call w_ε the right hand side of the above equality. From the definition of f_ε and Lemma 1(c) we have

$$\lim_{n\to\infty} \frac{u_\varepsilon(\rho_n\sigma) - w_\varepsilon(\rho_n\sigma)}{\gamma(\rho_n)} = 0 \text{ uniformly in } \sigma.$$

Let $\delta_n \equiv \operatorname{Max}|u_\varepsilon(\rho_n\sigma) - w_\varepsilon(\rho_n\sigma)|/\gamma(\rho_n)$. By the maximum principle applied to

$$u_\varepsilon(\rho\sigma) - w_\varepsilon(\rho\sigma) \text{ and } \delta_n\gamma(\rho)$$

on balls of radius ρ_n centered at the origin, we conclude that

$$u_\varepsilon(\rho\sigma) - w_\varepsilon(\rho\sigma) \equiv 0.$$

Let us note that from Lemmas 4 and 1(c) it follows that

$$u_\varepsilon(\rho\sigma)/\gamma(\rho) \to f_\varepsilon(\sigma) \text{ as } \rho \to \infty,$$

and hence that f_ε is independent of the sequence ρ_n used to define it.

It is not difficult to complete the proof of the theorem now if we realize that u_ε satisfies the harmonicity relation

$$u_{\varepsilon_1+\varepsilon_2}(\rho\sigma) = \int P_{\varepsilon_1}(\sigma\cdot\lambda) u_{\varepsilon_2}(\rho\lambda) \, dA(\lambda)$$

(since $P_{\varepsilon_1 + \varepsilon_2} = P_{\varepsilon_1} * P_{\varepsilon_2}$, where " $*$ " denotes the obvious "convolution" on the unit sphere), and hence, the same relation is satisfied by $f_\varepsilon(\lambda)$, making $g(r\sigma) = f_\varepsilon(\sigma)$ a positive harmonic function in the unit ball.

If we add to that the fact that

$$\int_{|\lambda| = 1} f_\varepsilon(\lambda) \, dA(\lambda) = \lim_{\rho_n \to \infty} \frac{1}{\gamma(\rho)} \int_{|\lambda| = 1} u_\varepsilon(\rho_n \lambda) \, dA(\lambda)$$

$$= \lim_{\rho_n \to \infty} \frac{1}{\gamma(\rho)} \int_{|\lambda| = 1} u(\rho_n \lambda) \, dA(\lambda) = C(u),$$

it follows that $f_\varepsilon(\sigma) = g(r\sigma)$ converges, for r going to 1, to a unique measure $\mu(S^{n-1})$ with total mass $C(u)$. The theorem now follows by letting $\varepsilon \to 0$ in the statement of Lemma 4.

2. FOURIER SERIES METHODS

The aim of this section is to use separation of variables and Fourier series techniques to study the asymptotic behavior of solutions to the equation

$$\Delta u - u = 0 \tag{1}$$

near infinity. In the first part of this section we shall show that as $|x| = r \to \infty$, the solution u, after being divided by a radial solution $\gamma(r)$ of the equation, approaches a generalized function $a(\omega)$ defined on the unit sphere. It corresponds to the "far field pattern" for the equation $\Delta u + u = 0$. In the second part of the section we shall show that for positive solutions of (1) $a(\omega)$ is indeed a measure, thus giving an alternate proof for Theorem 1. The essence of the proof of this last fact is an elementary lemma concerning Fourier series which seems to have other uses.

For the purpose of simplicity we shall first assume that the number of space dimensions is two and later indicate the required changes needed to make the arguments valid for general n. The main advantage of this is that we can give the arguments in terms of ordinary trigonometric (Fourier) series rather than the somewhat more cumbersome spherical harmonics. We begin by introducing the appropriate space of generalized functions on the unit circle in terms of Fourier series.

Let S be the space of formal Fourier series

$$a(\theta) \equiv \sum_{-\infty}^{\infty} c_{\nu} e^{in\theta} \tag{2}$$

(the c_{ν} complex constants) satisfying the condition: for every $\rho > 0$ there exists a positive constant $C(\rho)$ such that

$$|c_{\nu}| \leqslant \frac{C(\rho)|\nu|!}{\rho^{|\nu|}}. \tag{3}$$

By convergence in S we shall mean convergence of each coefficient separately, or, as we shall sometimes refer to it, *convergence by coefficients*. Elements of S need not be Fourier series of functions or even distributions. However, in case the series (2) does represent a distribution, we shall consider the series and the distribution the same object.

To every element $a(\theta)$ of S given by (2) we can associate a "smoothed out", "regularized", a "mollified" version $a^{t}(\theta)$ defined for $t > 0$ by the formula

$$a^{t}(\theta) = a(t, \theta) = \sum_{-\infty}^{\infty} c_{\nu} e^{-\nu^{2}t} e^{i\nu\theta}. \tag{4}$$

From Stirling's formula it follows that this series converges, can be differentiated indefinitely, and is a solution to the heat equation

$$a_{t}(t, \theta) = a_{\theta\theta}(t, \theta) \tag{5}$$

for positive t. Thus we may think of the space S as being contained in the (larger) space of boundary values (on $t = 0$) of solutions to the heat equation for $t > 0$.

THEOREM 2. (i) *The function u is a solution to*

$$\Delta u - u = 0 \tag{1}$$

in the whole plane, if and only if for every $r > 0$ it has a Fourier series given by

$$u(r, \theta) = \sum_{-\infty}^{\infty} c_\nu I_\nu(r) e^{i\nu\theta} \tag{6}$$

with coefficients c_ν satisfying (3).
  (ii) *If that is the case, the function*

$$u(r, \theta)/\gamma(r) \tag{7}$$

converges as $r \to \infty$ to an element $a(\theta)$ of S. Here $\gamma(r) = 2\pi I_0(r)$.
  (iii) *$a(\theta)$ uniquely determines the solution u, and can be chosen to be any element in S.*
  (iv) *We have the representation*

$$u(r, \theta) = \int_{0 \leqslant \phi \leqslant 2\pi} e^{x\cos\phi + y\sin\phi} a(\phi)\, d\phi \tag{8}$$

where this is to be interpreted as follows:

$$u^t(r, \theta) = \int e^{x\cos\phi + y\sin\phi} a^t(\phi)\, d\phi \text{ for all } t > 0. \tag{9}$$

$I_\nu(r)$ is the "Bessel function with imaginary argument" of order ν regular at zero:

$$I_\nu(r) = \sum_{m=0}^{\infty} \frac{\left(\tfrac{1}{2}r\right)^{\nu+2m}}{m!\,\Gamma(\nu+m+1)}. \tag{10}$$

(For integral $\nu \geqslant 0$, $I_{-\nu}(r) \equiv I_\nu(r)$.) In our discussion r will always be nonnegative and in the two-dimensional case ν will be integral. We note that $I_0(r)$ is positive for $r \geqslant 0$ and is a radial solution to equation (1) defined and regular in the whole plane.

Let us recall two basic estimates for the functions I_ν, for which we refer to Watson's treatise [9] for example. (See pages 17 and 203.)

$$I_\nu(r) \sim \frac{e^r}{(2\pi r)^{1/2}} \text{ for } r \to \infty, \nu \text{ fixed.} \tag{11}$$

$$I_\nu(r) \sim \left(\frac{1}{2}r\right)^\nu / \Gamma(\nu+1) \text{ as } \nu \to \infty \text{ for every fixed } r. \tag{12}$$

Now suppose u is a solution to equation (1). By standard methods we obtain the representation (after introducing polar coordinates)

$$u(r,\theta) = \sum_{-\infty}^{\infty} c_\nu I_\nu(r) e^{i\nu\theta}, \tag{6}$$

valid in the whole plane. From (11) we know that

$$I_\nu(r)/I_0(r) \to 1 \text{ as } r \to \infty \tag{13}$$

for ν fixed. It follows that the Fourier series of the function of θ

$$u(r,\theta)/\gamma(r) \tag{14}$$

converges by coefficients to

$$a(\theta) = \sum_{-\infty}^{\infty} c_\nu e^{i\nu\theta}. \tag{15}$$

Since the coefficients c_ν uniquely determine u, so does $a(\theta)$. We wish to show that $a(\theta)$ belongs to S.

Since for $r > 0$

$$\int |u(r,\theta)|^2 \, d\theta \tag{16}$$

is finite, the series

$$\sum_{-\infty}^{\infty} |I_\nu(r)|^2 |c_\nu|^2 \tag{17}$$

converges and its νth term approaches zero, hence is bounded as $|\nu| \to \infty$. Hence

$$I_\nu(r)|c_\nu| \leqslant M(r) \text{ for every } r. \tag{18}$$

Combining (18) (and the fact that it holds for every r) with (12) we obtain (3). Conversely it is easily seen that (3) insures that the series (6) converges, may be differentiated, and is a solution to (1) in the whole plane.

To see that (iv) holds, define

$$u_1^t(r, \theta) = \int_{0 \leqslant \phi \leqslant 2\pi} e^{x\cos\phi + y\sin\phi} a^t(\phi)\, d\phi. \tag{19}$$

By part (c) of Lemma 1 in Section 1, or the remarks following the statement of the theorem in Section 1, we know that since $a^t(\phi)$ is continuous (smooth, in fact)

$$u_1^t(r, \theta)/\gamma(r) \to a^t(\theta) \text{ as } r \to \infty. \tag{20}$$

The convergence is uniform in θ, in particular by coefficients. Since $a^t(\theta)$ uniquely determines u^t, we have $u_1^t = u^t$, for $t > 0$, hence $u_1 = u$.   Q.E.D.

Positive Solutions

LEMMA 5. *Suppose for each $r \geqslant r_0 > 0$ the series*

$$\sum_{-\infty}^{\infty} c_\nu(r) e^{i\nu\theta} \tag{21}$$

represents a nonnegative integrable function or measure $m_r(\theta)$, and, as $r \to \infty$, converges, by coefficients, to the series

$$\sum_{-\infty}^{\infty} c_\nu e^{i\nu\theta}. \tag{22}$$

Then the series (22) represents a nonnegative measure $m_\infty(\theta)$ and furthermore the norms $|m_r|$ remain uniformly bounded for $r_0 \leqslant r \leqslant \infty$.

Proof. We note that the nonnegativity of the measures m_r imply

$$|c_0(r)| = |m_r|,$$

and since $c_0(r)$ converges to c_0, the measures m_r are uniformly bounded for $r_0 \leqslant r < \infty$. It follows from a standard selection principle that there exists a sequence $r_i \to \infty$ such that the measures m_{r_i} converge to a nonnegative measure m. On the other hand, $c_\nu(r_i)$ converges, by hypothesis, as $i \to \infty$, to c_ν. Thus we may define $m_\infty = m$.

Alternate proof of Theorem 1. We have noticed that the function of θ given by (14) converges by coefficients to $a(\theta)$ given by (15). By the lemma $a(\theta)$ must be a nonnegative measure. The rest of the proof follows as in Section 1, or as in the proof of (iv) of Theorem 2.

Generalization to n dimensions. The contents of this whole section generalizes in a routine way to n dimensions. The essential step is the replacement of formula (6) by

$$u(r\omega) = \sqrt{2\pi} \sum_{\nu=0}^{\infty} \sum_{j=1}^{N_j} c_{\nu j} r^{(2-n)/2} I_{\nu+(n-2)/2}(r) K_{\nu j}(\omega), \quad (6)_n$$

where ω is a point on the unit sphere and $K_{\nu j}$ are the N_j linearly independent normalized spherical harmonics of degree ν. The role of the coefficients c_ν in the estimate (3) is now taken by

$$c_\nu \equiv \left(\sum_{j=1}^{N_j} |c_{\nu j}|^2 \right)^{1/2}.$$

The positive radial solution γ, (also denoted by γ in Section 1) which equals $2\pi I_0(r)$ in the two-dimensional case, is now the unique entire radial solution $\gamma(r)$ with $\gamma(0) = \omega_n$ (the area of the unit sphere in R^n). It is given by

$$\gamma(r) = Cr^{(2-n)/2}I_{(n-2)/2}(r),$$

with C chosen so that $\gamma(0) = \omega_n$. It is easily seen that Condition (3) remains unchanged. The representation (8) should, of course, be replaced by

$$u(x) = \int_{|\omega|=1} e^{x\cdot\omega}a(\omega)\,d\omega,$$

and similarly for (9).

Remarks concerning the lemma of this section. This lemma is of much wider applicability than stated here. For example it could apply to eigenfunction expansions for the Laplacian on a compact manifold. In that context it yields Widder's representation for positive solutions of the heat equation.

3. A HARNACK INEQUALITY

A corollary of Theorem 1 is the following estimate for the growth at infinity of positive solutions of the equation

$$\Delta u - u = 0.$$

COROLLARY. $u(x) \leqslant u(0)e^{|x|}$.

We want to point out that a similar estimate is true for positive solutions u of $\mathcal{Q}(u) - u = 0$, where $\mathcal{Q}(u) = a_{ij}D_{ij}u$ is uniformly elliptic, i.e. (using summation convention),

$$\lambda\Sigma\xi_i^2 \leqslant a_{ij}\xi_i\xi_j \leqslant \Lambda\Sigma\xi_i^2,$$

provided a local Harnack inequality holds: Denoting by $B_1(x)$ the unit ball centered at x,

$$\inf_{B_1(x)} u(y) \geqslant Cu(x).$$

(See [8], [3] for sufficient conditions.)

In such a case we can assert that for $x > 1$

$$u(x) \leq \frac{1}{\tilde{C}} u(0) |x|^{\alpha} \exp(\lambda^{-1/2} |x|) \qquad (*)$$

where α depends only on λ, Λ and the dimension, and where $\tilde{C} = \tilde{C}(\lambda)$.
The proof consists simply of comparing $u(y)$ with

$$g(y) = \tilde{C} u(x) |x - y|^{-\alpha} \exp(-\lambda^{-1/2} |y - x|)$$

outside of $B_1(x)$, and choosing α as to make g a subsolution. To be more precise, from the local Harnack inequality it follows that on the set $|x - y| = 1$

$$u(y) \geq g(y) \text{ with } \tilde{C} = C \exp(\lambda^{-1/2}).$$

Applying the maximum principle to the subsolution $\text{Max}\{g - u, 0\}$ over a sequence of large balls with $B_1(x)$ removed, it follows that $u(y) \geq g(y)$ outside $B_1(x)$. Hence if $|x| > 1$, (setting $y = 0$)

$$u(0) \geq C \exp(\lambda^{1/2}) u(x) |x|^{-\alpha} \exp(-\lambda^{-1/2} |x|),$$

proving ($*$).

It is interesting to point out that the rate of growth at infinity is independent of the constant in the local Harnack inequality, this entering only in the value of $\tilde{C}$.

REFERENCES

1. P. Hartman and C. Wilcox, "On solutions of the Helmholtz equation in exterior domains," *Math. Z.*, **75** (1961), 228–255.
2. A. Koranyi, "A survey of harmonic functions on symmetric spaces," *Proc. Symposia Pure Math. XXV*, part 1 (1979), 323–344.
3. N. V. Krylov and M. V. Saforov, *Math. USSR-Izv.*, **16**, no. 1 (1981), 151–164.
4. E. M. Landis, "Some problems in the qualitative theory of second order elliptic equations," *Russian Math. Surveys*, **18**, no. 1 (1963), 1–62.
5. C. Müller, "Über die ganzen Lösungen der Wellengleichung," *Math. Ann.*, **124** (1952), 235–264.

6. ———, "Ein neuer Zugang zur Theorie der Besselfunktionen in mehreren Dimensionen," *Ann. Acad. Sci. Fenn. Ser. A I Math.*, **415** (1968), 1–42.

7. H. Niemeyer, "Lokale und asymptotische Eigenschaften der Lösungen der Helmholtzschen Schwingungsgleichung," *Jahresber. Deutsch. Math.-Verein*, **65** (1962), 1–44.

8. J. Serrin, "On the Harnack inequality for linear elliptic equations," *J. Analyse Math.*, **4** (1955/56), 297–308.

9. G. N. Watson, *Theory of Bessel Functions*, 2nd ed., Cambridge Univ. Press, Cambridge, 1944.

June 1981

INDEXES

(BY AUTHOR)

265

JOHANNES C. C. NITSCHE, "MINIMAL SURFACES AND PARTIAL DIFFERENTIAL EQUATIONS" (PAGES 69–142)

STEVEN OREY, "PROBABILISTIC METHODS IN PARTIAL DIFFERENTIAL EQUATIONS" (PAGES 143–205)

JAMES RALSTON, "GAUSSIAN BEAMS AND THE PROPAGATION
OF SINGULARITIES" (PAGES 206–248)

LUIS A. CAFFARELLI AND WALTER LITTMAN,
"REPRESENTATION FORMULAS
FOR SOLUTIONS TO $\Delta u - u = 0$ IN R^{n}" (PAGES 249–263)